2nd International Conference and Young Scientist School on Magnetic Resonance Imaging in Biomedical Research 2015

AA003291

Journal of Physics: Conference Series Volume 677

Tomsk, Russia
7 – 9 September 2015

Editors:

A.V. Naumova
M.Y. Khodanovich
V.L. Yarnykh

ISBN: 978-1-5108-2088-3
ISSN: 1742-6588

Printed from e-media with permission by:

Curran Associates, Inc.
57 Morehouse Lane
Red Hook, NY 12571

proceedings.com

Some format issues inherent in the e-media version may also appear in this print version.

Printed by Curran Associates, Inc. (2016)

For permission requests, please contact the Institute of Physics
at the address below.

Institute of Physics
Dirac House, Temple Back
Bristol BS1 6BE UK

Phone: 44 1 17 929 7481
Fax: 44 1 17 920 0979

techtracking@iop.org

Additional copies of this publication are available from:

Curran Associates, Inc.
57 Morehouse Lane
Red Hook, NY 12571 USA
Phone: 845-758-0400
Fax: 845-758-2634
Email: curran@proceedings.com
Web: www.proceedings.com

Table of contents

Volume 677

2nd International Conference and Young Scientist School "Magnetic resonance imaging in biomedical research"
7–9 September 2015, Tomsk, Russia

Accepted papers received: 8 December 2015
Published online: 16 February 2016

Preface
011001

2nd International Conference and Young Scientist School "Magnetic resonance imaging in biomedical research" OPEN ACCESS A V Naumova, M Y Khodanovich and V L Yarnykh

011002

Peer review statement OPEN ACCESS

Papers

012001

Mathematical modeling of the magnetization transfer effect in tissues OPEN ACCESS V Yarnykh pg. 1

012002

New Image Reconstruction Methods for Accelerated Quantitative Parameter Mapping and Magnetic Resonance Angiography OPEN ACCESS J V Velikina and A A Samsonov pg. 8

012003

Reconstruction of unbroken vasculature of mouse by varying the slope of the scan plane in MRI OPEN ACCESS S Maltseva, A Cherevko, A Khe, A Akulov, A Savelov, A Tulupov, E Derevtsov, M Moshkin and A Chupakhin pg. 14

012004

Fluorocarbon compounds in MRI diagnostics and medical therapy OPEN ACCESS Yu Pirogov pg. 21

012005

The registration of signals from the nuclei other than protons at 0.5 T MRI scanner OPEN ACCESS N Anisimov, D Volkov, M Gulyaev, O Pavlova and Yu Pirogov pg. 25

012006

Demyelination of subcortical nuclei in multiple sclerosis OPEN ACCESS E Krutenkova, G Aitmagambetova, M Khodanovich, J Bowen, B Gangadharan, L Henson, A Mayadev, P Repovic, P Qian and V Yarnykh pg. 30

012007

MRI study of the cuprizone-induced mouse model of multiple sclerosis: demyelination is not found after co-treatment with polyprenols (long-chain isoprenoid alcohols) OPEN ACCESS M Khodanovich, V Glazacheva, E Pan, A Akulov, E Krutenkova, V Trusov and V Yarnykh pg. 32

012008

MR histology of advanced atherosclerotic lesions of ApoE- knockout mice OPEN ACCESS A Naumova, V Yarnykh, M Ferguson, M Rosenfeld and C Yuan pg. 38

012009

The use of innovative gadolinium-based contrast agent for MR-diagnosis of cancer in the experiment OPEN ACCESS V Chernov, A Medvedeva, I Sinilkin, R Zelchan, E Grigorev, I Frolova and I Nam pg. 41

012010

3D DCE-MRA of pedal arteries in patients with diabetes mellitus OPEN ACCESS M Zamyshevskaya, V Zavadovskaya, M Zorkaltsev, V Udodov and E Grigorev pg. 47

012011

Magnetic Resonance Imaging (MRI) with retrograde intralumen contrast enhancement of the rectum in diagnostics of rectovaginal fistulas after combination therapy of rectal cancer. Experience of application OPEN ACCESS A Usova, I Frolova, S Afanasev, A Tarasova and S Molchanov pg. 52

012012

The role of MRI in the diagnosis of acute radiation reaction in breast cancer patient
OPEN ACCESS Zh A Startseva, L I Musabaeva, AV Usova, I G Frolova, K A Simonov and V V Velikaya pg. 56

012013

Magnetic Resonance (MR) rectography in diagnostics of small-size rectal neoplasms
OPEN ACCESS AV Usova, IG Frolova, NG Trukhacheva, OV Cheremisina and SG Afanas'ev pg. 60

2nd International Conference and Young Scientist School "Magnetic resonance imaging in biomedical research"

A.V. Naumova [1,2], M.Y. Khodanovich[1,2], V.L.Yarnykh[1,2].

[1] *Department of Radiology University of Washington, Seattle, WA, USA*
[2] *Research Institute of Biology and Biophysics, National Research Tomsk State University, Tomsk, Russia.*

ABSTRACT

The Second International Conference and Young Scientist School "Magnetic resonance imaging in biomedical research" was held on the campus of the National Research Tomsk State University (Tomsk, Russia) on September 7-9, 2015.

The conference was focused on magnetic resonance imaging (MRI) applications for biomedical research. The main goal was to bring together basic scientists, clinical researchers and developers of new MRI techniques to bridge the gap between clinical/research needs and advanced technological solutions. The conference fostered research and development in basic and clinical MR science and its application to health care. It also had an educational purpose to promote understanding of cutting-edge MR developments. The conference provided an opportunity for researchers and clinicians to present their recent theoretical developments, practical applications, and to discuss unsolved problems.

The program of the conference was divided into three main topics. First day of the conference was devoted to educational lectures on the fundamentals of MRI physics and image acquisition/reconstruction techniques, including recent developments in quantitative MRI. The second day was focused on developments and applications of new contrast agents. Multinuclear and spectroscopic acquisitions as well as functional MRI were presented during the third day of the conference.

We would like to highlight the main developments presented at the conference and introduce the prominent speakers. The keynote speaker of the conference Dr. Vasily Yarnykh (University of Washington, Seattle, USA) presented a recently developed MRI method, macromolecular proton fraction (MPF) mapping, as a unique tool for modifying image contrast and a unique tool for quantification of the myelin content in neural tissues. Professor Yury Pirogov (Lomonosov Moscow State University) described development of new fluorocarbon compounds and applications for biomedicine. Drs. Julia Velikina and Alexey Samsonov (University of Wisconsin-Madison, USA) demonstrated new image reconstruction methods for accelerated quantitative parameter mapping and magnetic resonance angiography.

Finally, we would like to thank the scientific committee, the local organizing committee and the National Research Tomsk State University for giving an opportunity to share scientific ideas and new developments at the conference and the Russian Science Foundation (project № 14-45-00040) for financial support.

Journal of Physics: Conference Series **677** (2016) 011002 doi:10.1088/1742-6596/677/1/011002

Peer review statement

All papers published in this volume of *Journal of Physics: Conference Series* have been peer reviewed through processes administered by the proceedings Editors. Reviews were conducted by expert referees to the professional and scientific standards expected of a proceedings journal published by IOP Publishing.

Journal of Physics: Conference Series **677** (2016) 012001 doi:10.1088/1742-6596/677/1/012001

Mathematical modeling of the magnetization transfer effect in tissues

V Yarnykh [1,2]

[1] Vascular Imaging Lab, Department of Radiology, University of Washington, Seattle, WA, USA
[2] Neurobiology Lab, Research Institute of Biology and Biophysics, Tomsk State University, Tomsk, Russian Federation

E-mail: yarnykh@uw.edu

1. Biophysical mechanisms of proton magnetization exchange in biological objects

The term magnetization transfer (MT) is commonly used to define a group of phenomena caused by incoherent exchange of magnetic energy between water and macromolecules in biological objects. Historically, the first experimental evidence of MT was the observation of bi-exponential longitudinal relaxation in muscle and water-collagen mixtures [1]. Later, saturation transfer from macromolecules to water in tissues was discovered in experiments with off-resonance radiofrequency (RF) saturation [2], which gave rise to a narrower interpretation of MT as the effect specifically related to saturation transfer. In this lecture, the common nature of these observations is highlighted in view of a general mathematical model.

Understanding of magnetization transfer in tissues is closely related to an underlying physicochemical model of proton exchange and molecular motion. Considering a tissue as a mixture of water and biological macromolecules (for example, protein globules), four distinct compartments need to be taken into account: bulk water, hydration water at the water-polymer interface (possibly can be divided into subpopulations with different residence times), hydrophilic surface groups of a polymer molecule, and bulk polymer protons not directly interacting with water. The entire picture of possible pathways of magnetization exchange is rather complex including chemical exchange between water and various functional groups of a biopolymer (e.g. amide or hydroxyl), dipolar interactions, diffusion of water molecules, and spin diffusion within the macromolecule. More details about the compartmental structure of water-macromolecular systems of various degree of complexity can be found in Refs. [1, 3-13].

For practical purposes, the model can be simplified assuming that proton exchange involving bulk water, immobilized water on the water-polymer interface, and exchangeable groups of the polymer is fast in the NMR timescale. Also, it is usually assumed that spin diffusion within the protein molecule (similar to solids) results in fast equilibration of magnetic energy between surface and bulk protons of the polymer. Correspondingly, the model can be reduced to two proton fractions with distinct spectral properties: macromolecular protons with restricted motion characterized by a very wide line of ~20-30 kHz (typically invisible in the absorption spectrum but can be observed in saturation transfer spectra) and mobile water protons responsible for a directly observable signal with a narrow line of several Hz. Magnetic interaction between these two compartments is mostly enabled due to a relatively slow cross-relaxation process involving dipolar coupling (also termed as Nuclear Overhauser Effect, NOE) between hydration water or exchangeable protons (e.g. O-H, N-H, S-H) and adjacent non-exchangeable protons (e.g. C-H) of a biopolymer. If note, slow chemical exchange of mobile protons can contribute in a similar way and mathematically undistinguishable from cross-relaxation. This model is termed as the two-pool or binary spin-bath model, and the compartments have several synonymic definitions (free, mobile, or liquid pool for bulk water and bound, restricted, or semi-solid pool for macromolecular protons). Formally, magnetization exchange in the two-pool model is described by the effective first-order rate constant regardless of actually involved molecular mechanisms. The two-pool model provides a theoretical background for the majority of current studies of the MT effect in MRI, and it typically allows an adequate description of the observed MT phenomena within the accuracy of experimental data.

2. Two-pool model: mathematical description

Magnetization dynamics in the two-pool model is described by modified Bloch equations including cross-relaxation terms [14]:

$$\mathrm{d}M_x^{\mathrm{F}}/\mathrm{d}t = -M_x^{\mathrm{F}}/T_2^{\mathrm{F}} - 2\pi\Delta M_y^{\mathrm{F}} \tag{1a}$$

$$\mathrm{d}M_y^{\mathrm{F}}/\mathrm{d}t = 2\pi\Delta M_x^{\mathrm{F}} - M_y^{\mathrm{F}}/T_2^{\mathrm{F}} - \omega_1(t)M_z^{\mathrm{F}} \tag{1b}$$

$$\mathrm{d}M_z^{\mathrm{F}}/\mathrm{d}t = \omega_1(t)M_y^{\mathrm{F}} - (R_1^{\mathrm{F}} + k)M_z^{\mathrm{F}} + k(1-f)/f\, M_z^{\mathrm{B}} + R_1^{\mathrm{F}}(1-f)M_0 \tag{1c}$$

$$\mathrm{d}M_z^{\mathrm{B}}/\mathrm{d}t = -(R_1^{\mathrm{B}} + k(1-f)/f + \pi\omega_1^2(t)g^{\mathrm{B}}(\Delta, T_2^{\mathrm{B}}))M_z^{\mathrm{B}} + kM_z^{\mathrm{F}} + R_1^{\mathrm{B}} fM_0, \tag{1d}$$

where $M_{x,y,z}^{\mathrm{F,B}}$ are the x, y, and z-components of magnetization of the free (F) and bound (B) spins; Δ and ω1(t) are the frequency offset and amplitude of an RF pulse; $g^{\mathrm{B}}(\Delta, T_2^{\mathrm{B}})$ is the absorption line shape of bound spins; $R_1^{\mathrm{F,B}} = 1/T_1^{\mathrm{F,B}}$; k is the effective cross-relaxation rate constant for F→B transfer; f is the molar fraction of bound spins; and M0 is the equilibrium magnetization. The lineshape function gB was first introduced in the stationary solution of equations (1a-1d) [4] as outlined in Section 4. It was found that the standard Bloch approach resulting in the Lorentzian function gB demonstrated a disagreement with quantitative saturation transfer experiments, and alternative line shapes (Gaussian, Superlorenztian, Kubo-Tomita) were introduced in a semi-empirical way to achieve adequate description of experimental data [4, 6, 15]. This approach was then extended to the time-dependent formalism [14]. Several studies [6, 16, 17] have demonstrated that the most adequate model for tissues is achieved with the Superlorenztian function gB:

$$g^{\mathrm{B}}(\Delta, T_2^{\mathrm{F}}) = \sqrt{\frac{2}{\pi}} \int_0^{\pi/2} \frac{T_2^{\mathrm{B}}}{|3\cos^2\theta - 1|} \exp\left(-2\left(\frac{2\pi\Delta T_2^{\mathrm{B}}}{3\cos^2\theta - 1}\right)^2\right) \sin\theta \mathrm{d}\theta \tag{2}$$

It is important to consider several practically useful solutions of modified Bloch equations for the two-pool model. The presence of two coupled magnetization populations in equations (1a-1d) causes two key effects observed in experiments: bi-exponential longitudinal relaxation and saturation transfer. A combined action of these effects takes place in pulsed MT imaging experiments. A more detailed mathematical consideration and experimental applications of the two-pool model are given in next sections.

3. Relaxation in the absence of RF field

The free relaxation solution applies in the absence of the radiofrequency (RF) field. In this case, evolution of longitudinal components is independent of transverse components and has a bi-exponential form [1]:

$$M_z^{\mathrm{F}} = (1-f)M_0 + C_-\exp(-R_-t) + C_+\exp(-R_+t), \quad (3)$$

where

$$R_\pm = 1/2[R_1^{\mathrm{F}} + R_1^{\mathrm{B}} + k/f \pm \sqrt{(R_1^{\mathrm{B}} - R_1^{\mathrm{F}} + k/f - 2k)^2 + 4k^2(1-f)/f}\,] \quad (4)$$

and

$$C_\pm = \pm\frac{(R_1^{\mathrm{F}} - R_\mp)[M_{zi}^{\mathrm{F}} - M_0(1-f)] + k[M_{zi}^{\mathrm{F}} - M_{zi}^{\mathrm{B}}(1-f)/f]}{R_+ - R_-}, \quad (5)$$

where MziF,B are the initial values of longitudinal magnetizations, which are determined by an applied RF excitation technique. There are two decay modes described by "slow" (R-) and "fast" (R+) eigenvalues of the relaxation matrix R defined in Section 5. The "slow" eigenvalue defines the observed R1=1/T1 in the two-pool model: if R1F=R1B=R1, R-=R1 (exactly). The "fast" eigenvalue is on the order of the cross-relaxation rate constant defined for the bound-to-free pool transfer: assuming the same approximation (i.e. R1F=R1B=R1), R+=R1+k/f. Relative weights of decay modes depend on initial conditions (i.e. an RF excitation scheme) and the bound pool fraction. It should be understood that standard T1 measurement techniques applied to the two-pool system typically result in the bi-exponential dependence of the observed signal on a variable timing parameter (e.g. TR or TI), though the fast decay component can be observed only at short time intervals after excitation (<50-100 ms).

The bi-exponential relaxation equation (equation (3) can be used for estimation of cross-relaxation parameters k and f after some approximations, which take into account the effect of a particular pulse sequence on the initial values MziF,B. Several methods were suggested for this purpose based on different excitation schemes, such as semi-selective inversion (1,18,19), binomial excitation [18], double inversion [20], multiple inversions [21], stimulated echo preparation [22], and progressive off-resonance saturation [23].

4. Stationary RF saturation

Another important solution is for the stationary (continuous wave, CW) saturation regime, which presumes infinitely long RF irradiation (in practice during the time $>5T_1$). This solution provides an analytical equation for data analysis in Z-spectroscopy, where the parameters of the two-pool model can be extracted from dependences of the longitudinal magnetization on the offset frequency and power of RF field. The simplest form of CW solution can be obtained for high offset frequencies assuming negligible direct saturation of the free pool [24]. Currently, the commonly accepted approach for data analysis in Z-spectroscopy (4.6) is based on the full stationary solution of equations (1a-1d) with an arbitrary lineshape describing saturation of the bound pool (typically Superlorenztian for tissues [6]):

$$M_{\mathrm{ss}}^{\mathrm{F}}/M_0^{\mathrm{F}} = \frac{R_1^{\mathrm{F}}R_1^{\mathrm{B}} + R_1^{\mathrm{F}}k(1-f)/f + R_1^{\mathrm{B}}k + R_1^{\mathrm{F}}W^{\mathrm{B}}}{R_1^{\mathrm{F}}R_1^{\mathrm{B}} + R_1^{\mathrm{F}}k(1-f)/f + R_1^{\mathrm{B}}k + (R_1^{\mathrm{F}}+k)W^{\mathrm{B}} + (R_1^{\mathrm{B}} + k(1-f)/f)W^{\mathrm{F}} + W^{\mathrm{B}}W^{\mathrm{F}}}, \quad (6)$$

where $M_0^F=(1-f)M_0$, and

$$W^{F,B} = \pi\omega_1^2 g^{F,B}(\Delta, T_2^{F,B}) \tag{7}$$

are the saturation rates for the free and bound pools with gB given by equation (2) and Lorentzian function gF,

$$g^F(\Delta, T_2^F) = \frac{1}{\pi}\frac{T_2^F}{1+(2\pi\Delta T_2^F)^2} . \tag{8}$$

A modified version of equation (6) was introduced to accommodate pulsed saturation in MT imaging by using a CW-equivalent power [25, 26] delivered during the sequence repetition time (TR).

5. Periodic off-resonance saturation in a gradient-echo sequence: MT imaging

The above solutions are relatively simple, though they are generally not applicable to MT imaging experiments in MRI. In MT imaging, off-resonance saturation is delivered by relatively short RF pulses, and these pulses are incorporated in a pulse sequence with a relatively short TR. Correspondingly, the model should take into account both pulsed MT saturation and the effects of periodic excitation and relaxation produced by a sequence. A relatively simple approximated model was developed for a spoiled gradient-echo sequence with pulsed off-resonance MT saturation [17,27]. In this model, evolution of magnetization is analyzed separately during four time intervals comprising the sequence cycle: off-resonance saturation pulse (t_m), delay for spoiling gradient (t_s), readout pulse (t_p), and delay for signal readout and relaxation (t_r). Assuming that the pulsed steady state is established, the resulting equation can be written in matrix form as

$$\mathbf{M}_z = (\mathbf{I}-\mathbf{E}_s\mathbf{E}_m\mathbf{E}_r\mathbf{C})^{-1}\{[\mathbf{E}_s\mathbf{E}_m(\mathbf{I}-\mathbf{E}_r)+(\mathbf{I}-\mathbf{E}_s)]\mathbf{M}_{eq}+\mathbf{E}_s(\mathbf{I}-\mathbf{E}_m)\mathbf{M}_{ss}\} , \tag{9}$$

where $\mathbf{M}_z$ is the vector with components M_z^F and M_z^B corresponding to the longitudinal magnetization immediately before the excitation pulse; $\mathbf{M}_{eq}$ is the vector of equilibrium magnetization with elements $M_0(1-f)$ and $M_0 f$; $\mathbf{M}_{ss}$ is the vector of steady-state longitudinal magnetization for which the free pool component is given by equation (6); **I** is the unit matrix; the matrix term $\mathbf{E}_m=\exp((\mathbf{R}+\mathbf{W})t_m)$ describes off-resonance saturation by an RF pulse with duration t_m; the terms $\mathbf{E}_s=\exp(\mathbf{R}t_s)$ and $\mathbf{E}_r=\exp(\mathbf{R}t_r)$ describe relaxation during delays before (t_s) and after (t_r) an excitation RF pulse; and the diagonal matrix $\mathbf{C}=\mathrm{diag}(\cos\alpha, 1)$ corresponds to instant rotation of the magnetization M_z^F by an excitation pulse with the flip angle α. The matrices **R** and **W** are defined as follows:

$$\mathbf{R} = \begin{bmatrix} -R_1^F - k & k(1-f)/f \\ k & -R_1^B - k(1-f)/f \end{bmatrix}, \quad \mathbf{W} = -\,\mathrm{diag}(\langle W^F\rangle, \langle W^B\rangle),$$

where <WF,B> are the time-averaged saturation rates for the pools during the saturation pulse. Approximating RF power by its time-averaged value, <ω12>,

$$\langle\omega_1^2\rangle = 1/t_m \int_0^{t_m} \omega_1^2(t)\mathrm{d}t \tag{10}$$

the saturation rates can be expressed as

$$\langle W^{F,B}\rangle = \pi\langle\omega_1^2\rangle g^{F,B}(\Delta, T_2^{F,B}) , \tag{11}$$

where gF,B(Δ, T2F,B) are given by equations (2) and (8). Note that the stationary approximation for the transverse magnetization of the free pool presuming the Lorentzian function gF is limited to high offset frequencies where evolution of transverse components is sufficiently fast in the timescale of the applied RF irradiation (tm >>1/Δ).

Equation (9) provides a convenient model for data analysis in pulsed Z-spectroscopic experiments. This model with some variations related to the consideration of the transverse magnetization of the free pool was used in several studies [17, 27, 28] to determine cross-relaxation parameters from imaging data obtained at variable offset frequency and power of pulsed RF saturation.

6. MT imaging: simple agebraic description

Magnetization transfer ratio (MTR) is a widely used simple empiric quantitative measure of the MT effect in tissues. MTR is calculated as the percentage of a signal decrease due to off-resonance saturation: MTR=100(S_{ref}-S_{mt})/S_{ref}, where S_{mt} and S_{ref} are the signal intensities with and without saturation, respectively. Whole-brain MTR mapping and histogram analysis have gained significant popularity in clinical neuroimaging studies of diseases causing diffuse brain damage, such as multiple sclerosis. Based on the theory presented in the previous section, a simple yet accurate model describing the relationship between MTR and key parameters characterizing the two-pool model as well as basic parameters of a pulse sequence can be obtained. Equation (9) can be further simplified assuming that TR is short, and the first-order approximation can be applied to exponential terms [17]. The resulting equation for MTR can be expressed as

$$\mathrm{MTR} \approx \frac{kf\,(t_{\mathrm{m}}/\mathrm{TR})\langle W^{\mathrm{B}}\rangle}{k(R_1 - (1-f)\ln\cos(\alpha)/\mathrm{TR}) + f\,(t_{\mathrm{m}}/\mathrm{TR})\langle W^{\mathrm{B}}\rangle(R_1 + k - \ln\cos(\alpha)/\mathrm{TR})}. \quad (12)$$

Additionally, we assume in equation (12) that direct saturation of the free pool is negligible ($W^B >> W^F$), which requires frequency offsets Δ>1-2 kHz, and that $R_1^F=R_1^B=R_1$. Despite these approximations, equation (12) provides a close agreement with the full Bloch model given by equations (1a-1d) [17]. To understand basic trends affecting experimantal results in MT imaging, let us consider several agbraical consequences of varying pulse sequence parameters in equation (12).

5.1. Effect of the saturation rate <WB>.

Through equation (11), $<W^B>$ is increased with an increase of the saturation power and a decrease of the offset frequency. For simplicity, we assume that the excitation flip angle α in equation (12) is low. In the weak saturation regime, MTR~$fT_1<W^B>$, thus being sensitive to macromolecular fraction and independent of k. In the strong saturation regime, MTR approaches the limit where MTR is independent of f: MTR~$kT_1/(1+kT_1)$. In practice, this limit is unachievable due to the presence of direct saturation of the free pool and SAR limitations (for in vivo studies). At typical conditions of clinical MT imaging, both cross-relaxation parameters (k and f) provide comparable contributions into MTR.

5.2. Effect of the excitation flip angle α.

This effect is typically overlooked in literature, since the common practice of low-angle MTR measurements is driven by the goal of the maximization of the observed MTR. In fact, as seen from the above limiting conditions, at low flip angles MTR is weighted by both cross-relaxation parameters (f and k) and T1. Most pathological changes in tissues result in an increase of T1 and a decrease of f and k. Correspondingly, opposite trends in these parameters reduce pathological sensitivity of MTR. As seen from equation (12), the contribution of R1 in the denominator terms is reduced with an increase of α through its function $-\ln(\cos(\alpha))$. Thus, MT imaging with high excitation flip angles can provide more sensitivity to cross-relaxation parameters themselves, though the range of MTR values is reduced.

5.3. Effect of TR.
As seen from equation (12), the duty cycle of the saturation pulse, tm/TR plays a role of the scaling factor for the saturation rate WB. Consequently, a decrease of TR results in an increased MT effect and vice versa. Additionally, a shorter TR is helpful for increasing the term $-\ln(\cos(\alpha))/TR$ to reduce the effect of T1 weighting on MTR as described above.

7. Conclusions
The MT effect plays an important role in various areas of MRI as a tool for modifying image contrast, a source of quantitative parameters for tissue characterization, and a cause of systematic errors in quantitative imaging. This lecture provides an overview of biophysical mechanisms of MT in tissues, in-depth mathematical consideration of the widely used two-pool model of MT, and a summary of experimental methods used to study MT phenomena.

Acknowledgments
The study was support by Russian Science Foundation (project № 14-45-00040). The author acknowledge Tomsk State University Competitiveness Improvement Program for the preparation of the manuscript.

References
[1] Edzes H T, Samulski E T 1978 The measurement of cross-relaxation effects in the proton NMR spin-lattice relaxation of water in biological systems: hydrated collagen and muscle. J Magn Reson **31** 207-229
[2] Wolf S D, Balaban R S 1989 Magnetization transfer contrast (MTC) and tissue water proton relaxation in vivo. Magn Reson Med **10** 135-144
[3] Koenig S H, Brown R D 3rd, Ugolini R 1993 A unified view of relaxation in protein solutions and tissue, including hydration and magnetization transfer. Magn Reson Med **29** 77-83
[4] Henkelman R M, Huang X, Xiang Q S, Stanisz G J, Swanson S D, Bronskill M J 1993 Quantitative interpretation of magnetization transfer. Magn Reson Med **29** 759-766
[5] Kuwata K, Brooks D, Yang H, Schleich T 1994 Relaxation-matrix formalism for rotating-frame spin-lattice proton NMR relaxation and magnetization transfer in the presence of an off-resonance irradiation field. J Magn Reson B **104** 11-25
[6] Morrison C, Henkelman R M 1995 A model for magnetization transfer in tissues. Magn Reson Med **33** 475-482
[7] Adler R S, Swanson S D, Yeung H N 1996 A three-component model for magnetization transfer. Solution by projection-operator technique, and application to cartilage. J Magn Reson B **110** 1-8
[8] Li J G, Graham S J, Henkelman R M 1997 A flexible magnetization transfer line shape derived from tissue experimental data. Magn Reson Med **37** 866-871
[9] McLaughlin A C, Ye F Q, Pekar J J, Santha A K, Frank J A 1997 Effect of magnetization transfer on the measurement of cerebral blood flow using steady-state arterial spin tagging approaches: a theoretical investigation. Magn Reson Med **37** 501-510
[10] Gochberg D F, Kennan R P, Maryanski M J, Gore J C 1998 The role of specific side groups and pH in magnetization transfer in polymers. J Magn Reson **131** 191-198
[11] Stanisz G J, Kecojevic A, Bronskill M J, Henkelman R M 1999 Characterizing white matter with magnetization transfer and T2. Magn Reson Med **42** 1128–1136

[12] Ceckler T, Maneval J, Melkowits B 2001 Modeling magnetization transfer using a three-pool model and physically meaningful constraints on the fitting parameters. J Magn Reson **151** 9-27
[13] van Zijl P C, Zhou J, Mori N, Payen J F, Wilson D, Mori S 2003 Mechanism of magnetization transfer during on-resonance water saturation. A new approach to detect mobile proteins, peptides, and lipids. Magn Reson Med **49** 440-449
[14] Graham S J, Henkelman R M 1997 Understanding pulsed magnetization transfer. J Magn Reson Imaging **7** 903-912
[15] Iino M 1994 Transition from Lorentzian to Gaussian line shape of magnetization transfer spectrum in bovine serum albumin solutions. Magn Reson Med **32** 459-463
[16] Quesson B, Thiaudiere E, Delalande C, Dousset V, Chateil J F, Canioni P 1997 Magnetization transfer imaging in vivo of the rat brain at 4.7 T: interpretation using a binary spin-bath model with a superLorentzian lineshape. Magn Reson Med **38** 974-980
[17] Yarnykh V L 2002 Pulsed Z-spectroscopic imaging of cross-relaxation parameters in tissues for human MRI: theory and clinical applications. Magn Reson Med **47** 929-939
[18] Gochberg D F, Kennan R P, Gore J C 1997 Quantitative studies of magnetization transfer by selective excitation and T1 recovery. Magn Reson Med **38** 224-231
[19] Gochberg D F, Gore J C 2007 Quantitative magnetization transfer imaging via selective inversion recovery with short repetition times. Magn Reson Med **57** 437-441
[20] Morris G A, Freemont A J 1992 Direct observation of the magnetization exchange dynamics responsible for magnetization transfer contrast in human cartilage in vitro. Magn Reson Med **28** 97-104.
[21] Gochberg D F, Kennan R P, Robson M D, Gore J C 1999 Quantitative imaging of magnetization transfer using multiple selective pulses. Magn Reson Med **41** 1065-1072
[22] Ropele S, Seifert T, Enzinger C, Fazekas F 2003 Method for quantitative imaging of the macromolecular 1H fraction in tissues. Magn Reson Med **49** 864-871
[23] Helms G, Piringer A 2005 Simultaneous measurement of saturation and relaxation in human brain by repetitive magnetization transfer pulses. NMR Biomed **18** 44-50
[24] Grad J, Bryant R J 1990 Nuclear magnetic cross-relaxation spectroscopy. J Magn Reson **90** 1-8
[25] Ramani A, Dalton C, Miller D H, Tofts P S, Barker G J 2002 Precise estimate of fundamental in-vivo MT parameters in human brain in clinically feasible times. Magn Reson Imaging **20** 721-731
[26] Tozer D, Ramani A, Barker G J, Davies G R, Miller D H, Tofts P S 2003 Quantitative magnetization transfer mapping of bound protons in multiple sclerosis. Magn Reson Med **50** 83-91
[27] Sled J G, Pike G B 2001 Quantitative imaging of magnetization transfer exchange and relaxation properties in vivo using MRI. Magn Reson Med **46** 923-931
[28] Yarnykh V L, Yuan C 2004 Cross-relaxation imaging reveals detailed anatomy of white matter fiber tracts in the human brain. NeuroImage **23** 409-424

Journal of Physics: Conference Series **677** (2016) 012002 doi:10.1088/1742-6596/677/1/012002

New Image Reconstruction Methods for Accelerated Quantitative Parameter Mapping and Magnetic Resonance Angiography

J V Velikina[1], A A Samsonov[2,3]

[1]Department of [1]Medical Physics, University of Wisconsin-Madison, Madison, USA

[2] Department of Radiology, University of Wisconsin-Madison, Madison, USA

[3] Neurobiology Lab, Research Institute of Biology and Biophysics, Tomsk State University, Tomsk, Russian Federation

E-mail: alex0259@yandex.ru

Abstract: Advanced MRI techniques often require sampling in additional (non-spatial) dimensions such as time or parametric dimensions, which significantly elongate scan time. Our purpose was to develop novel iterative image reconstruction methods to reduce amount of acquired data in such applications using prior knowledge about signal in the extra dimensions. The efforts have been made to accelerate two applications, namely, time resolved contrast enhanced MR angiography and T1 mapping. Our result demonstrate that significant acceleration (up to 27x times) may be achieved using our proposed iterative reconstruction techniques.

1. Introduction

Magnetic Resonance Angiography (MRA) is an attractive non-invasive imaging choice that has shown utility in visualizing vasculature and blood haemodynamics. MRA is general term comprising a number of imaging techniques, including both single image and time-resolved imaging with or without the use of contrast agents. While many of MRA techniques may benefit from increased spatial and/or temporal resolution, time-resolved contrast-enhanced (CE) imaging often is in a particular need of acceleration. Indeed, in many diseases useful information can be gleaned from separate images of arterial and venous enhancement, hence rapid passage of contrast puts a bound on the acquisition time of a single frame, depending on application. For instance, such pathologies as intracranial arterial-venous malformations may require temporal resolution on the order of 0.5 - 1 s [1], while imaging of upper and lower extremities would benefit from 3-7 second frames [2]. At the same time, inherently low speed of MR acquisitions limits the number of k-space samples that can be acquired within a given time period. This duality creates a typical for MRA need for a tradeoff between spatial resolution/coverage on one hand and temporal resolution on the other hand. An attempt to maintain both high spatial and temporal resolution inevitably leads to the need to reconstruct images from incomplete data. Quantitative parameter mapping is another area of MRI that can often benefit from acceleration, albeit for different reasons. The need to acquire k-space data for different values of control parameters often makes total acquisition time too long, increasing probability of patient motion. Since conventional reconstruction from undersampled data produces aliased images and SNR degradation, advanced imaging techniques are needed to address this problem.

The required acceleration can be achieved on either acquisition side (e.g., by using efficient non-Cartesian trajectories [3-5] or specialized Cartesian acquisitions [6-8]) or reconstruction side (e.g. by using constrained reconstruction). Often, the two are developed jointly to maximize the benefits, as is done in the recently developed compressed sensing (CS) technique [9], which provide a way to reconstruct images that admit a sparse representation from incomplete incoherently sampled data.

2. Theory

3. The problem of reconstructing image series $\mathbf{f}$ is a system of linear equations $\mathbf{Ef} = \bar{\mathbf{b}}$, where $\mathbf{E}$ is the encoding matrix, and $\bar{\mathbf{b}}$ is the vector of measured k-space data for all time frames. If k-space data is incomplete, the above equation has infinitely many solutions. To isolate a single solution, prior information about the image series is invoked to constrain the solution. Standard CS approaches exploit sparsity of the image series after application of general transforms $\mathbf{D}$, such as wavelets or spatial/temporal gradients [10-12], implemented through minimization of a cost function: $\min_{\mathbf{f}}(\|\mathbf{Ef} - \mathbf{b}\|_2 + \lambda\|\mathbf{Df}\|_1)$ Equation (1), whose size is measured by lp norms ($\|x\|_p^p = \sum|x_n|^p$), with the balance between the data consistency term (first summand) and the penalty term (second summand) adjusted by the size of regularization parameter λ. Still, if the underlying image series is not sparsified efficiently by the transform $\mathbf{D}$, the reconstruction may be biased towards the assumed behavior or suffer from unresolved aliasing artifacts. Therefore, proper choice of the transform $\mathbf{D}$ is an important factor. In this work, we propose to consider two different types of transforms: a general transform $\mathbf{D}$ equal to the second discrete derivative in temporal dimension and a data-driven one obtained by performing principal component analysis (PCA) of the training data. The first type of transform relies on the smoothness of temporal/parametric curves, which can be efficiently sparsified by double differentiation [13]. The second approach, termed Model Consistency COndition (MOCCO) [14], relies on linearized representation of non-linear operators mapping between image space and space of MR parameters. Such linearization is obtained by applying PCA to compress a set of analytical curves derived from a theoretical signal model equation or a low-resolution estimate of a time series obtained from a more fully sampled k-space center. Similar to other principal component basis (PCB) techniques [15,16], MOCCO assumes that temporal behaviour of most pixels can be approximated well by a linear combination of several waveforms. However, it allows for deviations from the chosen model as follows. Let $(\boldsymbol{w}_k)$ be the set of chosen representative waveforms that span a linear subspace $\boldsymbol{W}$ of waveforms satisfying the model assumption. Let $\boldsymbol{S}$ be a synthesis operator mapping a set of coefficients (c_k) to a linear combination $\sum c_k \boldsymbol{w}_k \in \boldsymbol{W}$; and $\boldsymbol{S}^*$ be its adjoint analysis operator yielding coefficients of projection of an arbitrary waveform w onto $\boldsymbol{W}$, $\boldsymbol{S}^*: w \to (< w, \boldsymbol{w}_k >)$. If the model assumption is satisfied for a given pixel, then its temporal waveform w is "close" to $\boldsymbol{W}$, so $\boldsymbol{SS}^* w \approx w$. Thus, the choice of $\mathbf{D} = \boldsymbol{SS}^* - \mathbf{Id}$ provides requisite sparsity of the image series, while the use of l_1 norm provides robustness of the reconstruction to outliers, for instance, due to pathologies.

4. Methods

Fully sampled datasets were acquired using hybrid radial (in-plane)/Cartesian (through-plane) SPGR sequence (TR = 5.5 ms, BW = 125kHz, 20 slices, 8 coil receivers) for a set of 12 flip angles

FA=[2°,3°,4°,5°,6°,7°,8°,9°,10°,12°,14°,16°] in a single scan session on a 3T scanner (MR 750, GE Healthcare; Waukesha, WI). The acquired data for a single central slice was retrospectively undersampled to simulate an interleaved undersampled radial acquisition with an acceleration factor of 20. The twelve interleaves were scheduled according to bit-reverse ordering. The images were obtained from the undersampled data by solving equation (1) with $\mathbf{D} = \Delta^2$. For comparison purposes, we also reconstructed images from the undersampled data using gridding with density compensation and iterative SENSE. Also, we used a pair of ideal flip angles for this case (calculated to be 3° and 16°) and the number of projections that could be acquired within the same scan time (100 projections per image) to assess the improvement in T1 estimation accuracy for the given scan time. As a reference, images from fully sampled data for all 12 flip angles were reconstructed. T1 values were then estimated from each set of the reconstructed images using non-linear minimization [17].

MOCCO reconstruction was validated in a CE exam from an intracranial aneurysms patient conducted according to the IRB at our institution. The patient was scanned on a 3.0 T clinical scanner (Discovery TM MR750, GE Healthcare, Waukesha, WI) with an 8-channel head coil using a hybrid radial (in-plane)/Cartesian (through-plane) acquisition during a contrast injection. The scan parameters were TE/TR=1.5/4 ms, FA=25o, BW=125 kHz, 20 slices, voxel size 0.86x0.86x2 mm3. The data were reconstructed from 15 radial projections per slice per 1.2 s time frame (acceleration factor R=27) using iterative SENSE [18], strictly constraining PCB different number of principal components and the proposed method with both l_1-norm and l_2-norm in the penalty term. Reconstructed images were compared for image quality and temporal waveform fidelity.

5. Results

Results of R1 (1/ T1) estimation are given in figure 1. The proposed approach provided the best accuracy due to lower noise level, better artefact reduction and higher spatial resolution in the reconstructed images. The increase in the standard deviation measured in a region in white matter was measured to be 2.13 times for the proposed reconstruction versus 3.82 in the SENSE reconstruction from 2 ideal flip angles (both relative to the reference fully sampled data).

Images in figure 2 show limited maximum intensity projections of two early time frames for the four reconstruction techniques. Note that while all three constrained reconstruction techniques provide good spatial resolution and adequate SNR, the algorithms relying on l_2-norm regularization exhibit premature enhancement of some vessels. This observation is further confirmed by examining temporal waveforms of the aneurysm and its feeding artery in figure 3.

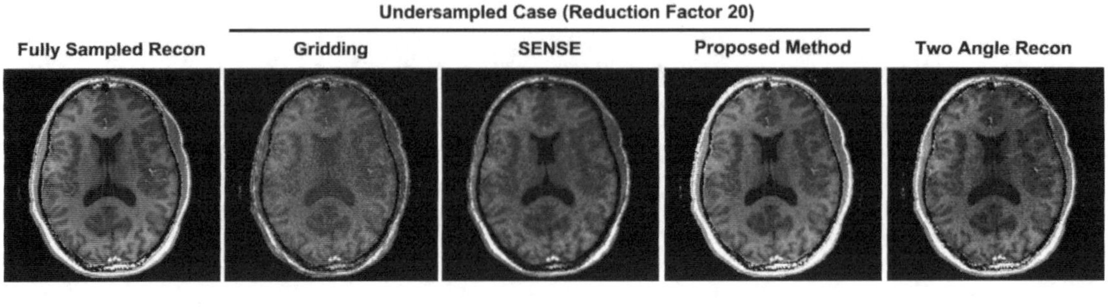

Figure 1. Reconstruction of T1 maps using different methods including proposed reconstruction and scan time equivalent two flip angle approach.

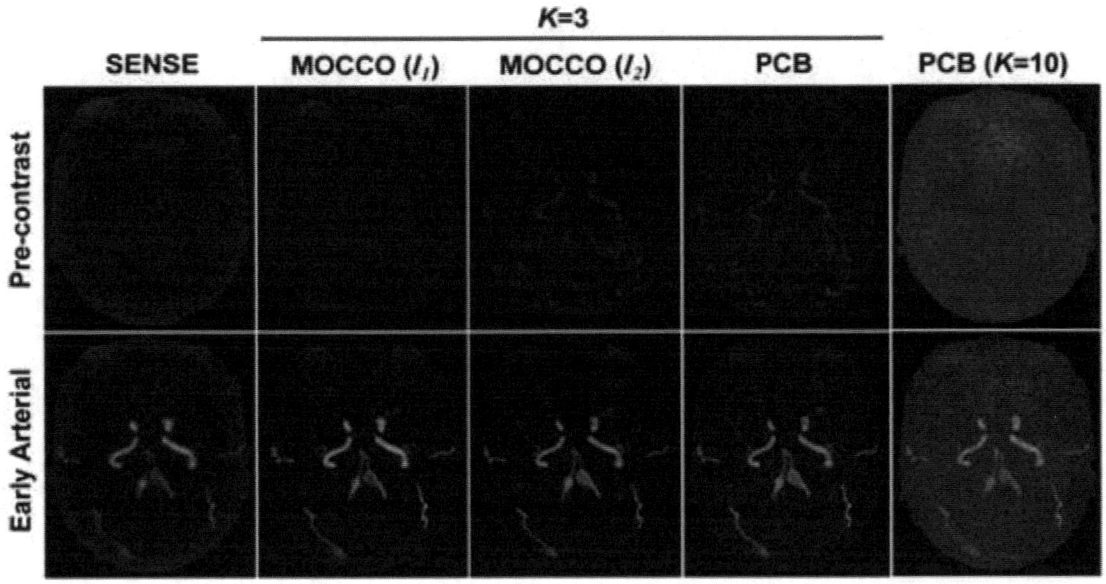

Figure 2. Time frames at 2.4 s (top row) and 7.2 s (bottom row) after the beginning of examination.

6. Conclusions

The proposed approaches exploiting sparsity of the image series in temporal/parametric dimension showed ability to improve image quality of accelerated acquisitions compared to parallel imaging/quadratic norm-based methods. In CE MRA applications, temporal dynamics in both normal and pathological vessels was preserved. In quantitative parameter mapping, the proposed method improved accuracy of parametric maps.

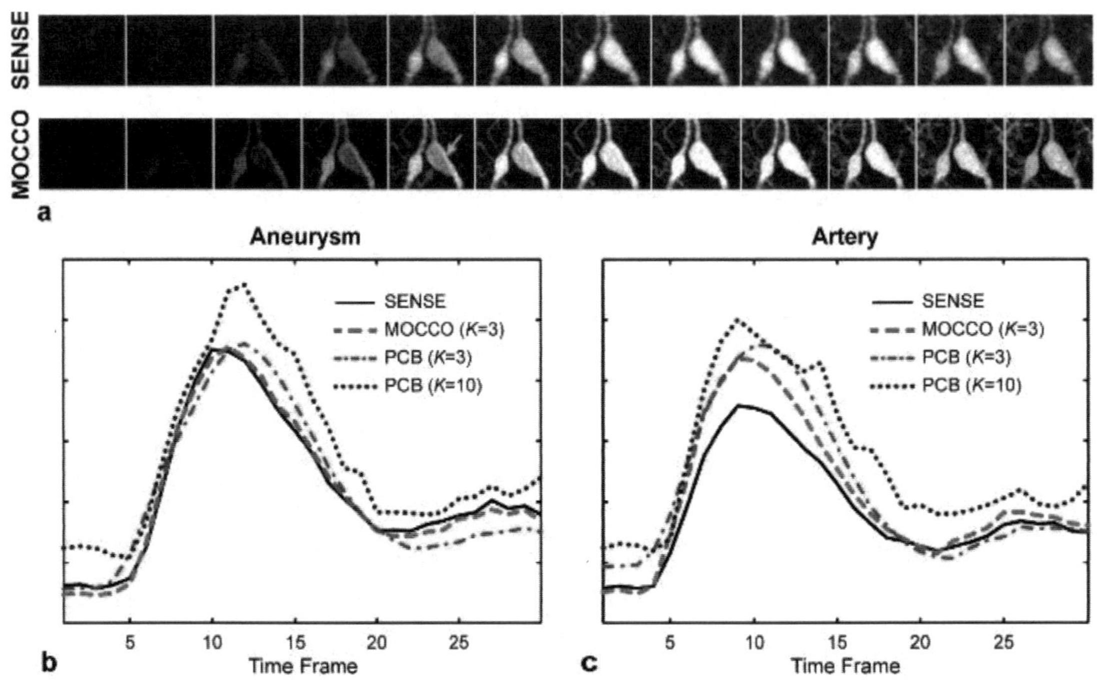

Figure 3. (a) Magnified ROI in the time series reconstructed by SENSE and MOCCO, (b, c): Contrast enhancement waveforms measured in the aneurysm and its feeding artery indicated by red and green arrows in (a), respectively.

Acknowledgement
The study was supported by Russian Science Foundation, project № 14-45-00040.

References:

[1] Hadizadeh D R, von Falkenhausen M, Gieseke J et al. 2008 Cerebral arteriovenous malformation:Spetzler-Martin classification at subsecond-temporal-resolution four-dimensional MR angiography compared with that at DSA. *Radiol* **246** 205–213

[2] Haider C R, Riederer S J, Borisch E A, et al. 2011 High temporal and spatial resolution 3D time-resolved contrast-enhanced magnetic resonance angiography of the hands and feet. *J. Magn. Reson. Imaging* **34** 2–12

[3] Peters D C, Grist T M, *Korosec* F R, et al. 2000 Undersampled projection reconstruction applied to MR angiography. *Magn Reson Med* **43** 91–101

[4] Meyer C H, Hu B S, Nishimura D G, Macovski A 1992 Fast spiral coronary artery imaging. *Magn Reson Med* **33** 202–213

[5] Barger A V, Block W F, Toropov Y, et al. 2002 Time-resolved contrast-enhanced imaging with isotropic resolution and broad coverage using an undersampled 3D projection trajectory. *Magn Reson Med* **48** 297–305

[6] Haider C R, Hu H H, Campeau N G, et al. 2008 3D high temporal and spatial resolution contrast-enhanced MR angiography of the whole brain. *Magn Reson Med* **60** 749–760

[7] Noll D C, Nishimura D G, Macovski A 1991 Homodyne Detection in Magnetic Resonance

Imaging, *IEEE Trans Med Imaging* **10** 154-163

[8] Muhammad U, Batchelor P G 2009 Optimized Sampling Patterns for Practical Compressed MRI" *SAMPTA'09, International Conference on Sampling Theory and Applications*

[9] Candès E J, Romberg J 2007 Sparsity and incoherence in compressive sampling. *Inverse Problems* **23** 969-985

[10] Block K T, Uecker M, Frahm J 2007 Undersampled radial MRI with multiple coils. Iterative image reconstruction using a total variation constraint. *Magn Reson Med* **57(6)** 1086-1098

[11] Lustig M, Donoho D, Pauly J M 2007 Sparse MRI: The application of compressed sensing for rapid MR imaging. *Magn Reson Med* **58(6)** 1182-1195

[12] Adluru G, Awate S P, Tasdizen T, Whitaker R T, Dibella E V 2007 Temporally constrainedreconstruction of dynamic cardiac perfusion MRI. *Magn Reson Med* **57(6)** 1027-1036

[13] Velikina J V, Alexander A L, Samsonov A A 2013 Accelerating MR parameter mapping using sparsity-promoting regularization in parametric dimension. *Magn Reson Med* **70(5)** 1263-1273

[14] Velikina J V, Samsonov A A Reconstruction of Dynamic Image Series from Undersampled MRI Data Using Data-Driven Model Consistency Condition (MOCCO). *Magn Reson Med*, early view

[15] Liang Z P 2007 Spatiotemporal imaging with partially separable functions. In: *Proc of ISBI; Washington, DC, USA* 988-991

[16] Pedersen H, Kozerke S, Ringgaard S, Nehrke K, Kim W Y 2009 k-t PCA: temporally constrained k-tBLAST reconstruction using principal component analysis. *Magn Reson Med* **62(3)** 706-716

[17] Deoni S C L, Rutt B K, Peters T M 2003 *Magn Reson Med* **49(3)** 515-526

[18] Pruessmann K P Weiger M, Bornert P, Boesiger P 2001 Advances in sensitivity encoding with arbitrary k-space trajectories. *Magn Reson Med* **46(4)** 638

Reconstruction of unbroken vasculature of mouse by varying the slope of the scan plane in MRI

S Maltseva[1,2,3], A Cherevko[2,3,4], A Khe[2,3,4], A Akulov[5,6], A Savelov[4], A Tulupov[3,4], E Derevtsov[1,3], M Moshkin[3,4,5], A Chupakhin[2,3,4]

[1]Sobolev Institute of Mathematics, Siberian Branch of the Russian Academy of Sciences, Novosibirsk, Russian Federation
[2]Lavrentyev Institute of Hydrodynamics, Siberian Branch of the Russian Academy of Sciences, Novosibirsk, Russian Federation
[3]Novosibirsk State University, Novosibirsk, Russian Federation
[4]International Tomography Center, Siberian Branch of the Russian Academy of Sciences, Novosibirsk, Russian Federation
[5]Cytology and Genetics Institute, Siberian Branch of the Russian Academy of Sciences, Novosibirsk, Russian Federation
[6] Neurobiology Lab, Research Institute of Biology and Biophysics, Tomsk State University, Tomsk, Russian Federation

E-mail: cherevko@mail.ru

Abstract. Reconstruction of vascular net of small laboratory animals from MRI data is associated with some problems. This paper proposes a method of MRI data processing which allows to eliminate the fragmentation of reconstructed vascular net. Problem of vessels fragmentation occurs in the case when vessels are parallel to the scanning plane. Our approach is based on multiple scanning, object under consideration is probed by several sets of parallel planes. The algorithm is applied to real MRI data of small laboratory animals and shows good results.

1. Introduction

In the conventional approach to reconstruction vasculatures based on MRA data, the object of interest is scanned by a series of parallel planes; the images of object slices acquired from the dataset. The imaging device registers the blood flow using the normal to section scan plane. The most informative sections are the ones with the most intense blood flow. Hence, the most informative scan, in terms of reconstructing vasculature, is the one where the set of the scan planes is perpendicular to the axes of the scanned vessels. In living organisms, however, it is impossible to choose one set of parallel planes that would be perpendicular to the axes of all blood vessels. This is due to both the various positions of the vessels and to vascular tortuosity. As a result, the reconstructed models of vasculatures contain interruptions in vessels that make these models unusable for hemodynamic studies.

The aforementioned reasons are characteristic of any study objects, but they are specifically common for studies in small laboratory animals.

Journal of Physics: Conference Series **677** (2016) 012003 doi:10.1088/1742-6596/677/1/012003

In terms of mathematics, the tomographic data $A(x, y, z)$ (image intensity) is dependence

$$A(x, y, z) = F(\vec{v}(x, y, z), \vec{n}) \quad (1)$$

from blood flow velocity vector $\vec{v}(x, y, z)$ at point (x, y, z) and normal to the scan plane vector $\vec{n}$. The function $F(\vec{v}(x, y, z), \vec{n})$ sets the image intensity value at (x, y, z) of a slice and takes small values at $(\vec{v}(x, y, z), \vec{n}) \approx 0$. There are two situations that correspond to the low image intensity in the images of slices:

1) There is no flow in the cross-section, or it is very slow: $\vec{v} \approx 0$

2) Velocity vector $\vec{v}$ and normal vector $\vec{n}$ are not collinear.

The first situation corresponds to the static tissue or thinning of a vessel (stenosis), while the second situation is caused by the specifics of the equipment and leads to the data loss. The paper deals with a technique that allows practitioners to overcome, to a certain extent, the problem of data loss, by varying the angle of the section view of a vessel at point (x, y, z).

From a mathematical point of view, the task of reconstructing the vasculature lies in defining the boundaries of a 3D area V in a way that

$$V = \{(x, y, z): A(x, y, z) > A_0\} \quad (2)$$

where $A_0 \in \mathbb{R}$ – is a numerical value that separates the scanned environment into blood vessels and adjacent tissues based on the intensity of images acquired.

2. Scan plane slope variation technique

Description of the principle of this method. A standard scan is carried out by using a single set of parallel planes, and, as a rule, their common normal vector is parallel to the axis of the scanned object. The technique of varying the slope of the scan plane lies in the deviation of the normal vector from its "standard position" by certain angles and scanning by using more than one set of planes designed by preprogrammed normal vectors. The number of different positions of the normal vector defines the number of scan planes sets and, hence, the number of acquired datasets.

The images of vessels will appear in different intensity depending on changed positions of the scan plane. The highest image intensity is achieved by the smallest deviation of the vessel's axis from the normal to the scan plane. The subsequent analysis of the acquired datasets allows for the reconstruction of the vasculature, practically, without interruptions.

Figure 1 schematically presents the images of the object sliced by planes with different normal vectors. Vector n_1 is the normal to one of the sets of parallel planes; one specimen of this set is shown in green (together with the cross-section of the study object). Vector n_2 is the normal to another set of parallel planes; one specimen of this set is shown in red (together with the cross-section of the study object).

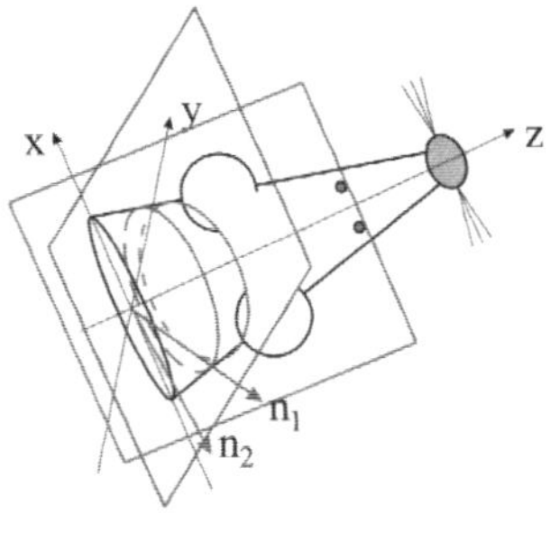

Figure 1. Cross-sections of the object made using different normal vectors

3. Steps of the vascularity reconstruction algorithm based on the method of varying the slope of the scan plane

3.1. Creating the area common for all datasets

The reasoning in this paragraph lies within the plane frame of the scanner. The authors used the imaging data recorded in DICOM format [17].

The data in each dataset is known in a 3D area in the form of a rectangular parallelepiped. The areas of different datasets are different. According to scanner specifications, each slice contains 256 rows and 256 columns. Let us express the parallelepiped containing the known data of dataset i as $\mathrm{K_i}$. Construct the smallest cube K, contains the $\cup_{\mathrm{i}=1}^{\mathrm{n}} \mathrm{K_i}$ set, that is, all parallelepipeds with known datasets.

Cube K is a "carrier" of all tomographic data at our disposal.

3.2. Interpolation of all datasets data onto a common 3D grid

Let us cover all side planes of cube K with a uniformly spaced 3D grid and find all datasets data approximation onto the 256×256×256 grid of cube K.

3.3. Combination of data from different datasets

Following the interpolation of data into the common grid area, we have a number (one array per each of datasets) of 3D arrays of the same size: B^i, $i = 1,\dots,n$, n – represents the number of available datasets. Having interpolated data from all datasets into the common grid area, we need to construct a single array containing all the information required.

3.4. Layer-by-layer pre-processing of data

MRA data has certain limitations that complicate its subsequent processing. Major limitations:

1. Inhomogeneous intensity of images (both of each given slice and across all of the slices). This effect is due to the inhomogeneity of the field in the receiver coil.
2. Layer-by-layer normalization of the reconstruction. In DICOM format, the image intensity of each given slice is scaled onto an 8- or 16-bit grid.
3. Data noise
4. Unclear boundaries of the blood flow.

These are the steps taken at the layer-by-layer data pre-processing stage.

1. Vertical alignment of the image intensity in a slice.
2. Horizontal alignment of the image intensity in a slice.
3. Adjustment of brightness and contrast settings
4. Noise reduction and image smoothing by applying the Fourier filter.

The layer-by-layer pre-processing stage allows for the reduction of some of the listed limitations.

3.5. Segmentation

For the purpose of subsequent hemodynamic calculations, the bitmap images acquired during the previous stages need to be converted into vector 3D format images, having highlighted the walls of the vessels (more precisely, the blood flow boundaries), that is, having segmented the images acquired. The segmentation is carried out in the ITK-Snap software environment [18].

3.6. Enhancement of fragmented vasculature. Local varying of segmentation parameters in small areas using the dialogue mode

As a rule, the vasculature acquired by applying the scan plane slope variation method with subsequent segmentation already contains a small number of fragments, but it may contain blood vessel interruptions. This occurs because the value set as the slicing parameter $\mathrm{A_0}$ in (2) of the previous stage cannot be uniformly used across the entire image. There are areas that visually contain vessels, however, the chosen slicing parameter is too high for the vessels in these areas to be segmented.

Journal of Physics: Conference Series **677** (2016) 012003 doi:10.1088/1742-6596/677/1/012003

For the purposes of segmentation, the parts of the vasculature containing interruption of a vessel are locked in subareas for which the slicing parameters are defined individually.

4. Realization of the proposed algorithm in MRI data of the mouse brain

The circulatory system of a laboratory mouse brain was chosen as an object of the study. Typically, mouse major vessels are 0.29 mm (vertebral artery). Experiments were carried out on unique scientific installation - Centre for Genetic Resources Laboratory Animals (RFMEFI61914X0005 and RFMEFI62114X0010). The field strength of the research scanner Bruker BioSpec 117/16USR used by the authors to acquire data presented in this paper - 11.7 T. For scanning, the "1H mouse brain surface coil" RF coil is used.

4.1. Tomography data details

Each data set contains 256 DICOM images of slices with associated metadata. All images are bitmaps in the shades of gray that are expressed by a 256 rows x 256 column square matrix. The reconstruction is carried out within cube *K* with an edge length of 24.9 mm.

4.2. Single dataset reconstruction

Reconstructing the mouse brain vasculature based on the data acquired by the standard scanning technique, the resulting vasculature following the main segmentation stage consists of 18 fragments. This vasculature is shown in figure 2.

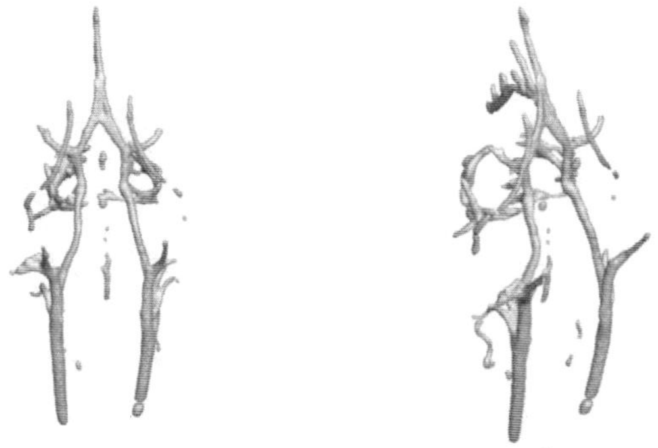

a b

Figure 2. Vasculature reconstructed based on the standard image acquisition technique in frontal view (a), turned at a 45- degree angle around longitudinal axis (b).

The resulting vasculature is marked by the pronounced fragmentation of vessels, and the missing fragments are quite large. This kind of vasculature cannot be used, neither in morphological studies, or in modelling the blood flow.

4.3. Reconstruction based on combined data from five datasets

Let us reconstruct the vasculature using the data acquired in the layer-by-layer pre-processing of the five datasets and their subsequent combination as the basis for segmentation. Following the segmentation, we acquire the vasculature consisting of 12 fragments (figure 3).

Notwithstanding the fact that this vasculature still lacks integrity, it is still more informative that above. We can see longer runs of vessels that have only local interruptions.

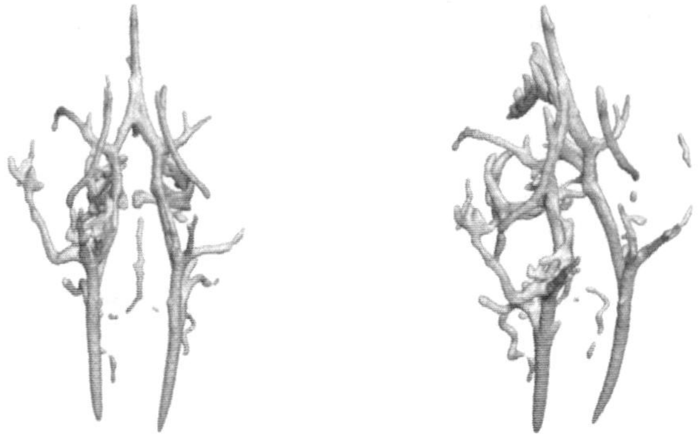

a b

Figure 3. Vasculature reconstructed based on the data from five datasets: in Willis's circle view (a); and as rotated (b).

4.4. Reconstruction based on the combined data from five datasets with subsequent enhancement using the dialogue mode

Having enhanced the vasculature described above (figure 3) using the dialogue mode, we have acquired the vasculature consisting of 2 fragments (figure 4).

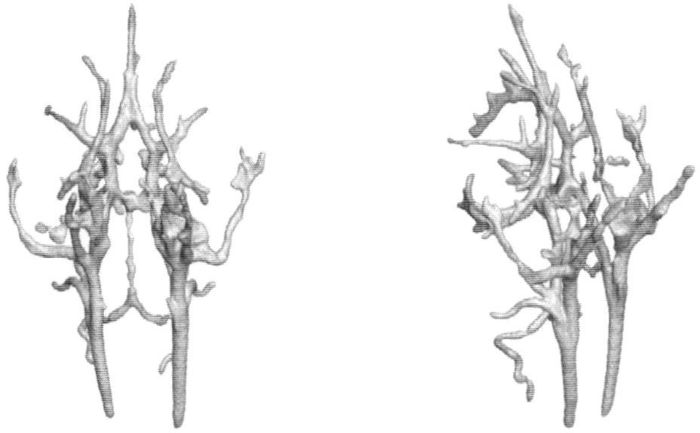

a b

Figure 4. The vasculature acquired based on the data from five datasets that was subsequently enhanced in the dialogue mode: in Willis' circle view (a), and as rotated (b).

This vasculature can be used for the modelling of the mouse brain blood flow.

5. Conclusion

The algorithm proposed allows for the reconstruction of the one-fragmental vascularity of mouse and rat brains. The resulting vasculature model as compared to the model reconstructed based on a single scan data, offers the following advantages:

1. It shows vessels that do not visualize in the model reconstructed based on a single scan.
2. It has less number of fragments than the single scan based model.

However, the acquired 3D model of the mouse brain vascularity has a number of drawbacks, and namely: somewhat increased diameter of the vessels; perivascular cerebrospinal fluid spaces are visualized.

The main disadvantage of our method is of course time consuming. The acquiring of each additional data package takes the same time as the standard one. Sure this fact decreases the method significance for medical applications. But in preclinical and biological researches (when time consuming is not so important) it may be successfully applied. Also our many-packages method require additional time as averaging and may replace that in some cases.

The task of reconstructing a volumetric vasculature of complex branched geometry is required in many applications (medicine, neurobiology); it is also a matter of interest for MRA practitioners. The currently available technique sallow for a limited realization of this task in terms of both its scope and quality. This task is closely related to the classic problem of vector tomography in reconstruction of a 3D vector based on the acquired imaging data. The proposed algorithm of varying the slope of the scan plane allows for the reconstruction of branched vasculatures and solves the problem of fragmented vessels in MRA data based models. The examples where the algorithm was used to reconstruct actual brain vasculatures in small laboratory animals, prove, beyond doubt, the efficiency of this method and demonstrate the associated advantages, and namely: significant decrease of fragmentation in the reconstructed vasculature and visualization of vessels that remain unseen using the standard reconstruction technique.

The advantages of the proposed algorithm give ground for its successful application in processing MRA data for its subsequent use in hemodynamic studies and phenotyping of genetic lines of laboratory animals with vascular pathologies.

Acknowledgements:
Work of A E Akulov, M P Moshkin, A A Savelov and A A Tulupov was completed thanks to the support of Russian Science Foundation (project #14-35-00020, all MRI experimentation studies using Bruker BioSpec 117/16 USR scanner). Work of A A Cherevko, A P Chupakhin, A K Khe was completed thanks to the support of Russian Foundation for Basic Research (project #14-01-00036, mathematical modelling). Work of S V Maltseva was completed thanks to the support of Russian Foundation for Basic Research (project #15-01-00745 A, mathematical modelling).

References

[1] Grist T M, Mistretta C A, Strother C M and Turski P A 2012 Time-resolved angiography: Past, present, and future *J. Magn. Reson. Imaging.* **36 (6)** 1273-86
[2] Beckmann N 2000 High resolution magnetic resonance angiography non-invasively reveals mouse strain differences in the cerebrovascular anatomy in vivo *Magn Reson Med* **44 (2)** 252-8
[3] Vorobtsova N A, Yanchenko A A, Cherevko A A, Chupakhin A P, Krivoshapkin A L, Orlov K Yu, Panarin V A and Baranov V I 2013 Modelling of cerebral aneurysm parameters under stent installation *RJNAMM* **28 (5)** 505-516
[4] Yanchenko A A, Cherevko A A, Chupakhin A P, Krivoshapkin A L and Orlov K Yu 2014 Nonstationary hemodynamics modelling in a cerebral aneurysm of a blood vessel *RJNAMM* **29 (5)** 307-317
[5] Miraux S, Franconi J M and Thiaudie`re E 2006 Blood Velocity Assessment Using 3D Bright-Blood Time Resolved Magnetic Resonance Angiography *Magn Reson Med* **56** 469-473
[6] Angenent S, Pichon E and Tennenbaum A 2006 Mathematical methods in medical image processing *Bulletin AMS* **43 (3)** 365-396
[7] Alvarez L, Guichard F, Lions P L and Morel J-M 1993 Axioms and fundamental equations of image processing *Arch. Rat. Mech. Anal.* **3** 199-257
[8] Alvarez L, Lions P L and Morel J-M 1992 Image selective smoothing and edge detection by nonlinear diffusion *SIAM J. Numer. Anal.* **29** 845-866
[9] Koenderink J J 1984 The structure of images *Biological Cybernetics* **50** 363-370
[10] Perona P and Malik J 1990 Scale-space and edge detection using anisotropic diffusion, *IEEE Trans. Pattern Anal. Mach. Intell.* **12** 629-639
[11] Roberts L 1965 Optical and electro-optical information processing, ch. Machine perception of

3-D solids. *MIT Press*
[12] Sobel I E 1970 *Camera models and machine perception* (Ph.D. thesis, Stanford Univ.)
[13] Witkin A, Kass M and Terzopoulos D 1987 Snakes: active contour models *Int. Journal of Computer Vision* **1** 321-331
[14] Casselles V, Catte F, Coll T and Dibos F 1993 A geometric model for active contours in image processing *Numerische Mathematic* **66** 1-31
[15] Sethian J, Malladi R and Vemuri B 1995 Shape modeling with front propagation: a level set approach *IEEE Trans. Pattern Anal. Mach. Intell.* **17** 158-175
[16] Morel J-M and Solimini S 1995 Variational methods in image segmentation *Birkhauser, Boston*
[17] http://medical.nema.org
[18] Yushkevich P A, Piven J, Hazlett H C, Smith R G, Ho S, Gee Js C and Gerig G 2006 User-guided 3D active contour segmentation of anatomical structures: Signi_cantly improved efficiency and reliability *Neuroimage* **31 (3)** 1116-28

Journal of Physics: Conference Series **677** (2016) 012004 doi:10.1088/1742-6596/677/1/012004

Fluorocarbon compounds in MRI diagnostics and medical therapy

Yu Pirogov

Department of Physics, Lomonosov Moscow State University, Moscow, Russian Federation

E-mail: yupi937@gmail.com

Abstract. The lecture describes the application of fluorocarbon compounds as blood substitutes and contrasting preparations in MRI diagnostics. A blood substitute product fluorocarbon Perfluorane® has shown effectiveness in oxygen delivery to the tissues of living organisms, and cardioprotective effect which does not depend on the patient's blood group. Inclusion of paramagnetic atoms (gadolinium, iron, etc.) to the Perfluorane® chemical formula creates a new compound with high MRI contrast efficiencies at Larmor frequencies of protons so and fluorine-19 nuclei.

1. Introduction

Magnetic resonance imaging (MRI), based on the effect of nuclear magnetic resonance (NMR), has become an integral part of the medical diagnosis. MRI scanners and NMR spectrometers are used widely in numerous biomedical applications. MRI can reveal different morphological changes of the internal organs and tissues. MRI signal in all medical imaging systems is formed on Larmor frequency of protons. It's not a coincidence that the protons as hydrogen nuclei in water molecules have a maximum concentration in most tissues of living organisms. This makes proton MRI suitable for high resolution imaging of majority of the organs.

2. Perfluorane® and its properties

There is virtually no fluorine atoms in a living organism (except for weak traces in the tooth enamel). Any injection of fluorine compounds respond to electromagnetic RF exposure of magnetic resonance scanner and can be detected when the excitation frequency of the electromagnetic wave is equal to the Larmor frequency of the fluorine nuclei. Fluorine-based MR images will have absolutely no background in normal tissues where fluorides are absent. Another reason to pursue development of NMR diagnosis on fluorine nuclei is the existence of fluorine-containing medicines, for example, blood substitute Perfluorane® [1,2]. It is critical to follow the distribution of the Perfluorane® after injection and to monitor the dynamics of its biochemical interactions with tissues.

Fluorine-based MRI can detect fluorine distrubution in the body after the injection, which helps to determine concentration of the oxygen carriers, retention in tissues and subsequent removal. A number of technical and fundamental issues have to be addressed for the successful application of MRI visualization of fluorine in medical practice. First of all, the transceiver transmit path of radio frequency channel has to be modified and adjusted it to the precession frequency of fluorine nuclei. Then, the imaging protocol has to be optimized in the order to form the most intensive fluorine images with high spatial resolution. Quantifiction of fluorine signal is possible with implementing localized fluorine nuclei NMR spectroscopy; this would allow determination of the metabolic profile of the specific tissue. Finally, fluorine images have to be co-localized with anatomical images obtained using proton MRI. Successful implementation of fluorine imaging and spectroscopy techniques is based primarily on the fact that the fluorine nuclei have very close characteristics of their magnetic parameters to protons, on which all known MRI techniques are built. In fact, the Larmor frequency of the fluorine nuclei is only 5% less than the respective frequencies of protons, natural content of fluorine-19 (^{19}F) is 100%, both nuclei have the same spin quantum number of ½ and close relative sensitivity.

Table 1 demonstrates that the fluorine-19 nucleus and protons have closest NMR characteristics compare to all other clinically interesting nuclei.

Table 1. NMR properties of clinically interesting nuclei.

Nucleus	Gyromagnetic ratio (MHz/T)	Spin quantum number	Relative content	Relative sensitivity
^{1}H	42.6	1\2	99	1.0
^{13}C	10.7	1\2	1.1	0.016
^{17}O	5.8	5\2	0.1	0.029
^{19}F	40.0	1\2	100	0.83
^{23}Na	11.3	3\2	100	0.093
^{31}P	17.2	1\2	100	0.07

Another reason for developing medical NMR fluoric techniques is a desire to create non-toxic fluorocarbon-based MRI contrast materials that can drastically improve imaging of internal organs and visualization of vascular network. Finally, a structure of fluorocarbon compounds in Perfluorane® may allow the use of Perfluorane® emulsion as bio-containers for the targeted delivery of pharmaceutical products to the zones of pathology.

It is known that fluorocarbon compounds can be present long time in the body without causing toxic effects and not altering organ functionality. This is a significant advantage in comparison with gadolinium-based MRI agents, which can cause toxic reactions. Perfluorocarbons (PFCs) are completely inert, not toxic, not metabolized in the body and excreted mainly with exhaled air.

Medical applications of fluoric compounds originate after discovering for fluorocarbon as an effective gas transporter. Note that fluorocarbons do not exist in nature and can be obtained by chemical substitution of all atoms of hydrogen by fluorine in hydrocarbons, therefore they are called sometimes "fluorocarbonhydrogens", although they do not contain hydrogen atoms. The emulsion of fluorocarbons in the water effectively absorbs oxygen and carbon dioxide, for example, liquid PFCs are able to dissolve more than 50% by volume oxygen and more than 150% carbon dioxide. The scientists of Nesmeyanov Institute of Organoelement Compounds RAS (INEOS RAS) [1.2] have made a major contribution to the development of this field, they discovered an outstanding gas transmission function of the PFC.

In 1966 year, Drs. Clark and Gollan [3] found that mouse submerged in the liquid PFC could long enough (up to 10 minutes) breathe the oxygen dissolved in PFC. Lungs of mouse were filled with liquid PFC, but in spite of this her breathing functions continued. After cleaning murine lungs from fluid and restoration the normal breathing process – the mouse continued to live, despite the prolonged period of deprivation from atmospheric oxygen.

3. Development of new fluorocarbon compounds for biomedicine

Immediately after this discovery, the United States, Japan and the USSR began work to create a substitute for blood plasma on the PFC basis. The joint research team of the Institute of Biophysics and INEOS RAS has developed recipes of Perfluorane®, as well as its manufacturing technology [2].

Perfluorane® is an aqueous micro-emulsion (droplet size 80-100 nm) mixture of two perfluoroorganic compounds, stabilized by polymeric surface-active substance (PSA). It quickly dissolves large amounts of oxygen, ensures the delivery of oxygen from the alveoli to erythrocytes and from erythrocytes to tissues, improves metabolism and gas exchange at the level of tissues, and also has a distinct cardio protective effect. Ministry of Health of the Russian Federation has allowed its use as blood substitute. Perfluorane® has saved the lives of hundreds of wounded soldiers who have experienced excessive blood loss during Afghanistan war. An important advantage of this artificial blood substitute is its indifference to the blood type.

PFCs are extremely resistant chemically and thermally, biologically stable and not metabolized in the body. PFCs are removed from the body mainly through the lungs with exhaled air. Speed of removal of these compounds is determined by two factors – the elasticity of PFC steam (boiling temperature) and its solubility in lipids, i.e. the ability to penetrate through membranes of pulmonary alveoli.

Fluorocarbon drug Perfluorane® is a mixture of two PFCs – perfluorodekalin and parametilcyklogeksilpiperidin, the first of which in itself is weak stable and quickly removed from the body, while the second one provides high stability of emulsion droplets in the mixture with greater duration of excretion from the body. The combination of these two compounds ensures stability and efficacy of the resulted emulsion, which can be stored in refrigerators up to 3 weeks and frozen for up to two years. It is important to outline that detailed studies have not shown any pathological changes in organs and tissues after complete removal of perfluorocarbons in spite of the long duration of their presence in the body.

High concentrations of the Perfluorane® in living tissues open up the opportunity to use this agent for medical diagnosis by detecting fluorine-19 signal with NMR (19F-MRI). Several different research groups in Netherlands, USA, and Japan applied 19F-MRI on artificial objects (phantoms) and laboratory animals and received high-intensity 19F-NMR three-dimensional images [4-6]. However, due to the fact that none of these countries did not have the drug type Perfluorane®, passed the full cycle of clinical research and allowed for medical use, these fragmented results are still very far from clinical applications.

The Moscow State University team leaded by academician Alexey R. Khokhlov continues development of the Perfluorane®-based MRI contrasting compounds and new reagents for targeted delivery of medicines to the damaged tissues [7-9].

Acknowledgements
These works are supported by the grant from the Ministry of Education and Science of the Russian Federation No. 14.604.21.0060 (RFMEF160414X0060).

References

[1] Moroz V V, Krylov N L, Ivanitsky G R 1995 *Anaesthesiology and Rheumatology* 6 pp 12-17 (*in Russian*)
[2] Maevsky E L. *et al.* 2003 Emulsion of perfluoroorganic compounds for medical purposes, a process for the preparation thereof and methods for treating and preventing diseases with the use thereof *US Patent* No. 6 **562** 872 B **1**
[3] Clark L S and Gollan F 1966 *Science* **152** 1755
[4] Sloviter H A and Kamimoto T *Nature* 1967 **216** 458
[5] Geyer R P, Monroe R G and Taylor K 1968 *Organ perfusion and preservation* ed J Norman, J Folkman, W Hardison, L Rudolf, F Veith (New York: Appleton-Century-Crofts) p 85
[6] Riess J G and Krafft M P 2006 *Blood Substitutes* ed R Winslow (Amsterdam Boston Heidelberg London New York Oxford Paris San Diego San Francisco Singapore Sydney Tokyo: Elsevier Academic Press) chapter 24 pp 259-275
[7] Gulyaev M et al 2013 *Radioelectronics(electronic journal)* 8 (http://jre.cplire.ru/jre/aug13/8/text.html)

[8] Volkov D *et al* 2014 *Radio electronics (electronic journal)* 11 (http://jre.cplire.ru/jre/nov14/11/text.html)
[9] Gulyaev M et al 2014 *Radio electronics (electronic journal)* 1 (http://jre.cplire.ru/jre/jan14/1/text.html)

The registration of signals from the nuclei other than protons at 0.5 T MRI scanner

N Anisimov[1], D Volkov[2], M Gulyaev[1], O Pavlova[2] and Yu Pirogov[2]

[1]Faculty of Fundamental Medicine, Lomonosov Moscow State University, Moscow, Russian Federation

[2]Faculty of Physics, Lomonosov Moscow State University, Moscow, Russian Federation

E-mail: anisimovnv@mail.ru

Abstract. The practical aspects of the adaptation of the medical MRI scanner for multinuclear applications are considered. Examples of high resolution NMR spectra for nuclei ^{19}F, ^{31}P, ^{23}Na, ^{11}B, ^{13}C, ^{2}H, and also NQR spectrum for ^{35}Cl are given. Possibilities of MRI for nuclei ^{19}F, ^{31}P, ^{23}Na, ^{11}B are shown. Experiments on registration of signals ^{19}F from the fluorocarbons injected in laboratory animals are described.

1. Introduction. Purpose

MRI is widely used for medical diagnostics. Besides, it is applied and to the solution of more wide range of tasks – technological applications, for example, MRI microscopy, and also to researches of small laboratory animals. Efficiency of researches significantly depends on opportunity to work not only at protons, but also other magnetic nuclei. Therefore it is desirable that in a complete set of the scanner there were the receiving coils (or transmitting/receiving ones) capable to be adjusted on the demanded NMR frequency and take place as far as possible close to the studied object. Producers of medical MRI scanners usually try to take away the users from temptations to use the equipment for the purposes not connected with medical diagnostics. Therefore technical obstacles for connection of not firm coils are created, access for the user to technical documentation and program opportunities are limited, the underestimated technical characteristics are declared.

However in some cases it is possible to bypass these restrictions and to expand operational opportunities of the scanner. We will show how it was succeeded to solve these problems for the standard medical scanner Bruker Tomikon S50. Thanks to it, it was succeeded to make not only demonstration experiments on an NMR and MRI for heavy nuclei, but also registration of signals ^{19}F from the fluorocarbons injected in laboratory animals [1].

2. Materials and methods

The transmitting/receiving (T/R) section of the scanner contains the T/R coil which is built in a magnet cavity. This one is the main coil. Further in the text for it the term "main transmitting coil" will be also used. It is connected either to the receiver, or to the transmitter by means of the passive T/R switch. Besides, in the scanner there are removable receiving coils adapted for research of separate human organs. They are connected to the receiver by means of similar T/R switches. The choice of the coil connected to the receiver is defined by the selector on which inputs signals from outputs of switches arrive. Output of the selector is connected to a receiver input. As switches firm blocks named as HPHP X-BB-Filterbox 12-32 MHz are used. It is obvious that the name of the block reflects its pass-band. As the transmitter (the radio-frequency (RF) field generator) the LPPA 2120 amplifier is used. It amplifies RF pulses of low power which are created on amplitude, duration, etc. It is natural to assume that in its name the top and lower borders of a pass-band (21 and 20 MHz) are also designated. Especially, as the similar parameters are declared in the firm documentation (Doc. No.: T4J-1133). Use of the amplifier and T/R switches with the above parameters is quite justified as it is supposed that the scanner is intended for registration only of protons for which Larmor frequency in the field of 0.5 T is equal to 21.08 MHz.

At first sight, the registration of the nuclei other than protons is impossible on this scanner. It concerns also fluorine ^{19}F which resonant frequency - 19.83 MHz only for 6 percent differs from the NMR proton frequency. However the need for registration of this nucleus forced to carry out the detailed analysis of technical characteristics of all T/R components. We took measurements which revealed that the amplitude-frequency response (AFR) of the LPPA 2120 is uniform in very wide range - from 8 and, at least, to 30 MHz and above. On a site from 8 to 6 MHz it smoothly decreases twice then remains almost uniform up to 3 MHz. At lower frequencies it sharply falls down - for 2 MHz almost by 10 times. Thus information on AFR of the amplifier is inadequately reflected in firm documentation. There are strong reasons to believe that the LPPA 2120 is a modification of the known LPPA 10010 (Dressler) model. Both models have not only similar AFR, but also visually similar designs. And modification is made to increase its output power – to 2 kW instead of 1 kW as in LPPA 10010.

We defined a set of the nuclei which are of interest to biology which in principle are possible for registration on our scanner taking into account AFR of RF generator. First of all, nuclei whose NMR frequencies lie in the range from 3 MHz and above as for this frequency range AFR falls down no more than twice. Nuclei were chosen: ^{19}F, ^{31}P, ^{11}B, ^{23}Na, ^{13}C and ^{2}H. Their Larmor frequencies in the field of 0.5 T are equal respectively 19.83, 8.53, 6.76, 5.73, 5.30 and 3.72 MHz.

We measured AFR of transmitting section of scanner taking into account passing of RF pulses not only through the amplifier, but the T/R switch. As opposed to the amplifier the AFR of the switch falls down at frequencies below 10 MHz more sharply. As a result at registration of nuclei ^{19}F, ^{31}P, ^{11}B, ^{23}Na,^{13}C and ^{2}H amplitude of impulses at the T/R switch output is attenuated respectively in 1, 3, 7, 14, 21 and 75 times.

Results of measurements specify that besides protons, only registration of fluorine ^{19}F is possible without serious problems. For registration of other nuclei it is necessary to increase amplitude of RF field. The simplest way for this purpose is to reduce the size of the transmitting coil. It, in turn, determines the size of the studied object. Therefore we were limited to registration of signals from samples volume ~10 cm^3 placed in cylindrical vessels with diameter ~2 cm.

For registration of NMR signals the ring coils of the small size (diameter of 2-4 cm) containing some rounds of a copper wire with a diameter of 0.5 mm were made. The coil was a part of the oscillatory contour containing also two trimmers - for resonance tuning and matching of resistance of a contour and input resistance of the T/R switch. This contour was connected to this switch instead of the main transmitting coil. But even at the small size of the coil not for all nuclei it was succeeded to solve a problem of generation of pulses with desirable amplitude completely. Additional problem was the matching of resistances - the input resistances of the T/R switch on the one hand and output resistances of the T/R coil and RF generator on the other hand. We noticed that the frequency is lower, the matching is worse. Therefore at registration of NMR signals of the nuclei other than protons and fluorine, it was necessary to look for a compromise. Namely, one can set short RF pulses and work at small flip angle or provide a large flip angle (~90 degrees) by means of increase in pulse duration.

In experiments on registration of the nuclei other than protons the experience of optimization of a T/R section was gained. The need for such optimization came to light in attempt to adapt this section for registration of ^{19}F NMR - high resolution spectra and MR images. We found out that firm receiving coils are capable to be adjusted not only on the proton NMR frequency (21.08 MHz) but also on the fluorine one (19.83 MHz). However the range of tuning of the main transmitting coil (20.80-21.31 MHz) was insufficient. Therefore two solutions of this problem were planned:

1) RF field amplification by means of the additional coil which is tuned on 19.83 MHz and it is inductive coupled with the main transmitting coil. This coil is capable to generate secondary RF field which amplitude is more primary which is formed by the main coil by QK times, where Q is tuned-circuit Q-factor, and K is coefficient of mutual induction [2-3]. As $Q \sim 10^2$ and $K \sim 10^{-1}$, amplification can rich some tens and be sufficient for providing a necessary angle of magnetization vector rotation (flip angle) by the acceptable amplitude and duration of RF pulse.

2) Use as the T/R coil of one of firm receiving coils after its modification. In this case the main transmitting coil is not used at all.

We tested both options. In both cases as the receiving coil the firm coil intended for study of a knee was used. The choice of such coil is caused by that its design is better suitable for placement in it a small laboratory animal – a rat. This coil is quadrature. It contains two receiving channels with coils which axes are mutually perpendicular. The channel with the coil whose axis is parallel to a main transmitting coil axis was modified.

Modification was as follows.

1) The voltage-controlled switch (with use a pin-diode) was installed between channel output and T/R switch input. It is necessary that Q-factor of coil was maximum for the period of action of RF pulse.

2) The voltage-controlled capacitors (varicaps) were replaced with ordinary trimmers. It is necessary as this channel in both options is used for RF power transfer.

3. Results

Having added to a firm configuration of the MRI scanner only homemade LC-tanks containing ring coils and trimmers we had an opportunity to register NMR signals not only from protons, but also other nuclei. In particular we registered ^{19}F, ^{31}P, ^{11}B, ^{23}Na, ^{13}C and ^{2}H.

For all nuclei, except ^{13}C and ^{2}H it was succeeded to receive also MRI. Images of the acceptable quality with the resolution of 2 mm were received after scanning by gradient echo (low flip angle) method in time about 5 minutes.

Besides, by means of the homemade probe adjusted on the frequency of 28 MHz and placed out of a scanner magnet the signal of a ^{35}Cl nuclear quadruple resonance (NQR) from potassium chlorate $KClO_3$ successfully was registered.

Examples of NMR spectra and MRI are given in figure 1. Curves are supplied with either a frequency scale or information on spectrum width (SW) in ppm. Spectral line widths at half height in Hertz are given also.

It should be noted that only for ^{2}H nucleus the isotope enriched sample is used. Other registered isotopes are presented in used samples at the natural abundance.

Registration of signals^{19}F was successfully carried out with use of both options. However the preference was given to the second one as in the first option of RF generator worked at the power close to maximum for it. It limited possibilities of a variation of RF pulse parameters (amplitude and duration). Besides, there was a risk of unstable work or even an overheating of the equipment at long experiments. These are factors it was considered when planning experiments with laboratory animals which duration could exceed one hour.

In figure 2 ^{19}F NMR spectra of synthetic fluorocarbon substances – Perfluoranum® and its main components are given [4]. Also the ^{19}F MRI of a rat for the second day after an intravenous injection of Perfluoranum® is presented. ^{1}H and a combination of ^{1}H and ^{19}F MRI are given for reference of anatomical structures [5].

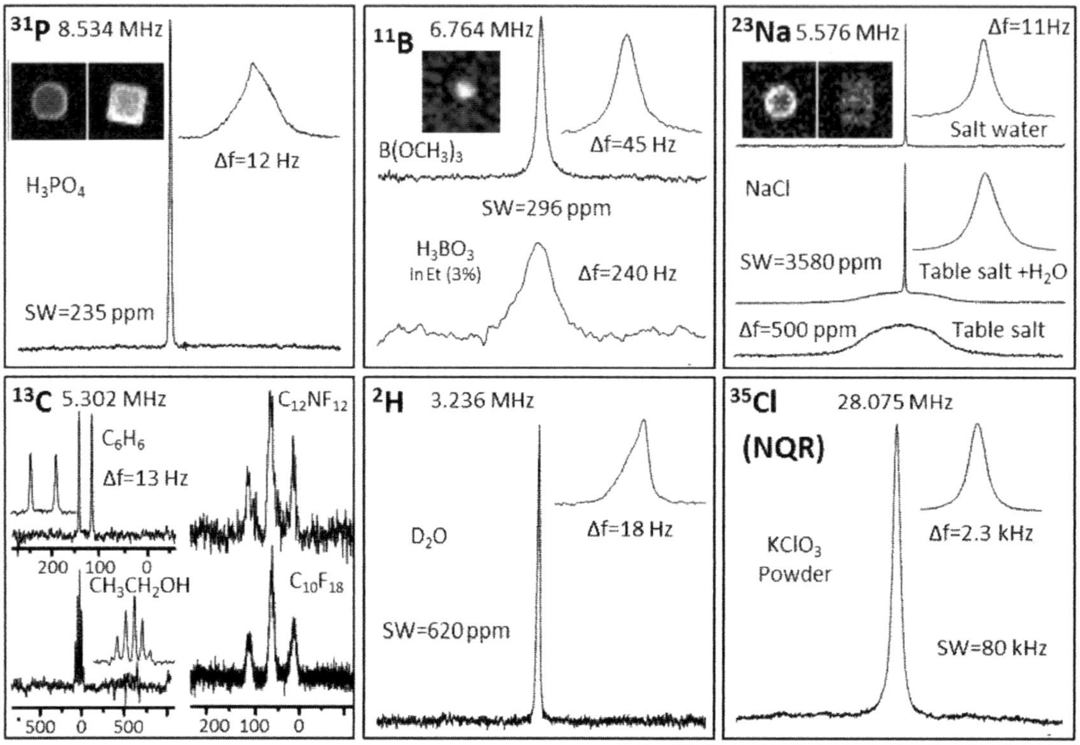

Figure 1. Examples of NMR spectra of nuclei other than proton received at 0.5 T MRI scanner.

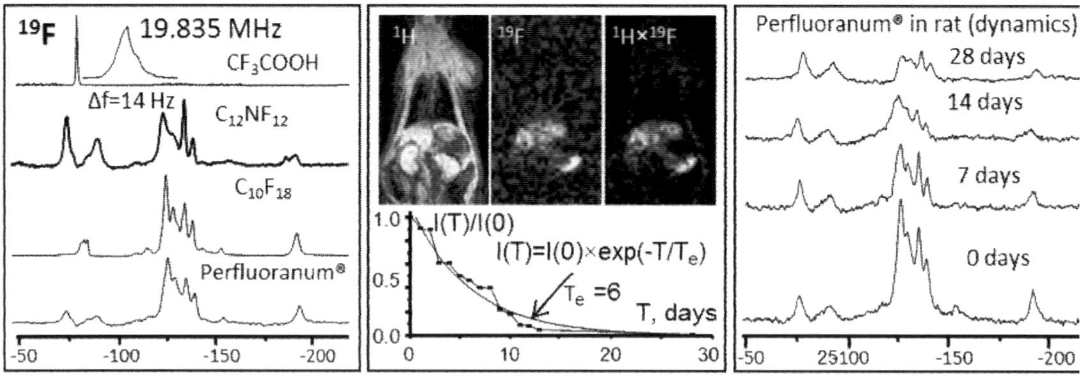

Figure 2.Left: ^{19}F NMR spectra of fluorocarbons. Center upper: MRI of rat for the second day after an intravenous injection of Perfluoranum®. Center bottom: curve of decay of perfluorodecaline in rat liver. Solid line is the simulation of decay with T_e=6 days. Right: ^{19}F spectra of whole body of rat at different days after Perfluoranum® injection.

4. Discussion. Conclusions

Efficiency of registration of heavy nuclei on our scanner can be increased if to replace firm T/R switch on similar, but with more suitable AFR. Other way is to work without this switch using the system containing the receiving coil and transmitting one with mutually perpendicular axes (cross-coil system). We do not plan to do it on ours of 0.5 T the scanner yet. It is not simple work. In general it is more preferable to solve similar tasks on high field (3 T and higher) the equipment. Eventually, the task was set not in at any cost to receive NMR spectra and MRI from heavy nuclei. The task consisted in as much as possible to develope operational opportunities of the equipment, having made only the minimum changes to its initial configuration.

As for registration of NMR signals ^{19}F how our practice showed the experience acquired on the low field scanner is useful and at statement of similar experiments on the high field equipment. And use of data from both scanners (in our case - 0.5 and 7 T) allows to specify ambiguities of measurements and finally to increase efficiency of researches. [5-6]

It is useful to carry out own technical analysis of the standard MRI equipment for an assessment of its true operational opportunities. Having such information it is possible to carry out useful improvements at which changes of a configuration remain minimum. Acquaintance with circuitry solutions of a concrete NMR equipment promotes more rational use of its resources and efficiency of the scientific research conducted on it.

Acknowledgements
This work was supported by the grant from the Ministry of Education and Science of the Russian Federation No. 14.604.21.0060 (RFMEF160414X0060).

References

[1] Jesu's Ruiz-Cabelloa, Barnett B P, Bottomley P A and Bulte J W M 2011 *NMR in Biomed.* **24** 114
[2] Flack F C, James E and Schlapp D 1971 *Med & Biol. Engng.* **9** 79
[3] Schnall M, Barlow C, Subramanian V and Leigh L Jr 1986 *J. Magn. Reson.* **68** 161
[4] Maevsky E I and Gervits L L, 2008 *Supplement of Chimica Oggi/CHEMISTRY TODAY, Focus on Fluorine Chemistry* **26**(3) 8
[5] Gulyaev M V, Gervits L L, Ustynyuk Yu A, Anisimov N V, Pirogov and Yu A Chochlov A R 2013 *Journal of radio electronics* No 8 http://jre.cplire.ru/alt/aug13/11/text.html (*in Russian*)
[6] Volkov D V, Pavlova O S, Gulyaev M V., Anisimov N V and Pirogov Yu A 2014 *Journal of radio electronics* No 11 - http://jre.cplire.ru/jre/nov14/7/text.pdf (*in Russian*)

Demyelination of subcortical nuclei in multiple sclerosis

E Krutenkova[1], G Aitmagambetova[1], M Khodanovich[1], J Bowen[2, 3], B Gangadharan [3], L Henson[3], A Mayadev[3], P Repovic[3], P Qian[3] and V Yarnykh[1,2]

[1] Neurobiology Lab, Research Institute of Biology and Biophysics, Tomsk State University, Tomsk, Russian Federation
[2]Vascular Imaging Lab, Department of Radiology, University of Washington, Seattle, USA
[3]Multiple Sclerosis Center, Swedish Neuroscience Institute, Seattle, WA, USA

e-mail: len—k@yandex.ru

Abstract. Myelin containing in basal ganglia in multiple sclerosis patients was evaluated using new noninvasive quantitative MRI method fast whole brain macromolecular proton fraction mapping. Myelin level in globus pallidus and putamen significantly decreased in multiple sclerosis patients as compared with healthy control subjects but not in substantia nigra and caudate nucleus.
Key words: multiple sclerosis, macromolecular proton fraction, demyelination, magnetic resonance imaging, subcortical gray matter.

1. Introduction
Multiple sclerosis was used to determine as white matter disease because plaques, main marker of MS, were clearly visualized [1]. At present it was obvious that gray matter damage which was poorly detectable using conventional MRI method occurred at the first stage of MS and correlated with clinical symptoms (both motor and cognitive) [2]–[4] Subcortical gray matter involvement in MS pathology was also observed [2], [5]but still need to be studied. In our research, basal ganglia demyelination in MS patients was evaluated using fast mapping of macromolecular proton fraction (MPF). The method of fast MPF mapping was developed by Yarnykh [6] and based on the magnetization transfer effect. Of note, relaxation-based MRI methods [7]–[9] are highly sensitive to paramagnetic ions and hardly appropriate for the assessment of demyelination in structures with high iron content, such as basal ganglia In contrast, MPF mapping is a highly-accurate quantitative method insensitive to iron and, therefore, very suitable for studying demyelination in gray matter in MS.

We present the results of secondary analysis of earlier published data [4] aimed to identify the effect of demyelination on MPF in subcortical gray matter structures in MS.

2. Methods
3D MPF maps were obtained using 3T MRI scanner from 30 MS patients and 17 healthy control subjects. Detailed characteristics of the study population can be found elsewhere [4]. Caudate nucleus, putamen, globus pallidus and substantia nigra were outlined on MPF maps. Mean MPF values in each structure were calculated as average measurements within contours weighted by the contour area and

then averaged between hemispheres. MPF values in each structure were compared between healthy control participants and MS patients using independent samples t-test ($p<0.05$).

3. Results
MPF values in the globus pallidus and putamen were significantly lower in the MS group as compared to the healthy control group ($p=0.008$ and $p=0.004$ respectively). The group differences between MPF in caudate nucleus and substantia nigra were not significant, though the trend of MPF decrease was observed (Table 1).

Table 1. Mean value of MPF in subcortical structures in control and MS groups

Structure	MS group	Control group	p
Caudate nucleus	6.997	7.131	0.238
Globus pallidus	9.656	10.150	0.001*
Putamen	7.129	7.508	0.001*
Substantia nigra	8.906	9.137	0.124

* Statistically significant changes of MPF value in subcortical structures in MS patients in compared with healthy volunteers.

In conclusion, subcortical structures in MS had different degree of myelin loss. The most demyelinated structures were the globus pallidus and putamen, but substantia nigra and caudate nucleus were minimally affected.

Acknowledgments
The study was supported by Russian Science Foundation (project №14-45-00040). The authors acknowledge Tomsk State University Competitiveness Improvement Program for the preparation of the manuscript.

References
[1] Hulst H and Geurts J 2011 *BMC Neurol.* **11** 153.
[2] Neema M, Arora A., Healy B, Guss Z, Brass S, Duan Y, Buckle G, Glanz B, Stazzone L, Khoury S, Weiner H, Guttmann C and Bakshi R 2009 *J. Neuroimaging* **19** 3–8.
[3] Modica C, Zivadinov R, Dwyer M, Bergsland N, Weeks R and Benedict R 2014 *AJNR. Am. J. Neuroradiol* 1–6.
[4] Yarnykh V, Bowen J and Henson L 2015 Radiology **274** 210-220
[5] Tjoa C, Benedict R, Weinstock-Guttman B, Fabiano A, Bakshi R, 2005 *J. Neurol. Sci.*, **234** 17–24.
[6] Yarnykh V. 2012 *Magn. Reson. Med.*, **68** 166–178 .
[7] Whittall K, MacKay A, Graeb D, Nugent R, Li D and Paty D 1997 *Magn Reson Med* **37** 34–43
[8] Deoni S, Rutt B, Arun T, Pierpaoli C and Jones D 2008 *Magn Reson Med* **60** 1372–1387.
[9] Hwang D, Kim D-H and Du Y 2010 *Neuroimage* **52** 198–204.

MRI study of the cuprizone-induced mouse model of multiple sclerosis: demyelination is not found after co-treatment with polyprenols (long-chain isoprenoid alcohols)

M Khodanovich[1], V Glazacheva[1], E Pan[1], A Akulov[1,2], E Krutenkova[1], V Trusov[3], V Yarnykh[1,4]

[1] Neurobiology Lab, Research Institute of Biology and Biophysics, Tomsk State University, Tomsk, Russian Federation
[2]Solagran Limited, Biotechnology Company, South Melbourne, Australia
[3] Institute of Cytology and Genetics, Siberian Branch of the Russian Academy of Sciences, Novosibirsk, Russian Federation
[4]Vascular Imaging Lab, Department of Radiology, University of Washington, Seattle, WA, USA

E-mail: khodanovich@mail.tsu.ru

Abstract. Multiple sclerosis is a neurological disorder with poorly understood pathogenic mechanisms and a lack of effective therapies. Therefore, the search for new MS treatments remains very important. This study was performed on a commonly used cuprizone animal model of multiple sclerosis. It evaluated the effect of a plant-derived substance called Ropren® (containing approximately 95% polyprenols or long-chain isoprenoid alcohols) on cuprizone-induced demyelination. The study was performed on 27 eight-week old male CD-1 mice. To induce demyelination mice were fed 0.5% cuprizone in the standard diet for 10 weeks. Ropren® was administered in one daily intraperitoneal injection (12mg/kg), beginning on the 6th week of the experiment. On the 11th week, the corpus callosum in the brain was evaluated in all animals using magnetic resonance imaging with an 11.7 T animal scanner using T2-weighted sequence. Cuprizone treatment successfully induced the model of demyelination with a significant decrease in the size of the corpus callosum compared with the control group ($p<0.01$). Mice treated with both cuprizone and Ropren® did not exhibit demyelination in the corpus callosum ($p<0.01$). This shows the positive effect of polyprenols on cuprizone-induced demyelination in mice.

1. Introduction

Multiple sclerosis (MS) is a neurological disorder of high social impact with poorly understood pathogenic mechanisms and a lack of effective therapies. For this reason, the search for new MS treatments is highly important. This study was carried out on a commonly used animal model of MS that is produced by administration of the neurotoxic agent cuprizone (CPZ). CPZ causes oligodendrocyte death followed by demyelination.

The CPZ model is a toxicological model that allows the study of acute and chronic demyelination caused by a primary loss of oligodendrocytes followed by spontaneous remyelination in the absence of inflammation [1]. This model of acute and chronic demyelination and spontaneous remyelination is simple to induce. The differentiation of inflammation, macrophage infiltration and demyelination in

MS lesions is important [2], and CPZ treatment allows the study of demyelination independent of inflammation. This model does not allow the study of autoimmune mechanisms of MS, but it is highly suitable to elucidate cellular and molecular mechanisms of demyelination and remyelination independent of peripheral immune system activity [3].

Ropren® is isolated from conifer needles and is mainly composed of polyprenols (long-chain isoprenoid alcohols) that have hepatoprotective and immunomodulatory effects, have the ability to restore function after liver damage and have efficacy as a treatment for patients with chronic alcoholism [4]. The functional activity of this substance is extensively studied but it is clear that it is of great interest for pharmacology [5]. Polyprenols and their derivatives are essential components of cellular membranes, affect the membrane transport of proteins or peptides, contribute to nervous impulse transmission, and are also involved in the biosynthesis of glycoproteins and cholesterol [6]. This suggests that polyprenols (long-chain isoprenoid alcohols) could have a positive effect on myelin loss in pathological conditions, such as MS, and this study examines the effect of Ropren® on CPZ-induced demyelination.

2. Methods

2.4 Animals and treatment

The rationale, design, methods and animal ethics for this study were approved by Ethics Committee of Biological Institute of Tomsk State University.

Twenty-seven eight-week-old male CD-1 mice were obtained from the vivarium of the Institute of Pharmacology of the Siberian Branch of the Russian Academy of Science for use in this study. Mice were housed at 21 ± 2 °C, humidity 40% ± 2%, 12/12 h light/dark cycle, food and water was provided ad libitum.

After 10 days of quarantine, the animals were divided into three groups: "Control", "Demyelination", and "Demyelination+Ropren". The CPZ animal model of MS was induced as previously described [1, 7]. The "Demyelination" and "Demyelination+Ropren" groups were fed the standard chow diet containing 0.5% of CPZ (Bis(cyclohexanone)oxaldihydrazone, Sigma-Aldrich) for 10 weeks. The "Control" group was fed the standard chow diet for 10 weeks. The animals in the "Demyelination+Ropren" group were injected by Ropren® (Solagran Limited) at a dose of 12 mg\kg intraperitoneally once a day from the 6th week of CPZ treatment. The animals of the other groups were injected with the same volume of the vehicle (vegetable oil).

2.2 *Ropren® substance*

Ropren® substance (95% polyprenols or long-chain isoprenoid alcohols) was isolated from the green verdure of *Picea abies* (L.) Karst as previously described [8]. Ropren® was supplied by (Solagran Limited).

2.5 2.3 MRI aqusition

On the 11th week, all animals underwent brain MRI examination with an 11.7 T small animal imager (Bruker BioSpec 117/16USR) with horizontal core. Imaging was performed under isoflurane anesthesia (1.5%–2% in oxygen) with respiratory monitoring (SA Instruments, Stony Brook, NY, USA) during the scan. The mice were placed on a heated bed (30°C) in order to maintain normal body temperature. The animals were placed in the prone position and were then slid into the magnet bore on an animal bed.

T2 weighted volume acquisition RARE sequence was used for in vivo MRI acquisition (spin echo-based RARE sequence, TR = 3500 ms, TE = 33 ms, RARE factor = 8, matrix: 224×224×80, slice thickness of 0.5 mm).

2.6 Image processing
Image analysis was performed using ImageJ software. Regions-of-interest (ROIs) corresponding to the corpus callosum were outlined manually (Figure 1). The size of the corpus callosum was evaluated as a sum of the area in three adjacent cross-sections (−0.82 ÷ −0.94 mm from bregma according to [8]). Delineation of corpus callosum was performed using a blind method, without the operator knowing the treatment group to which the animals belonged.

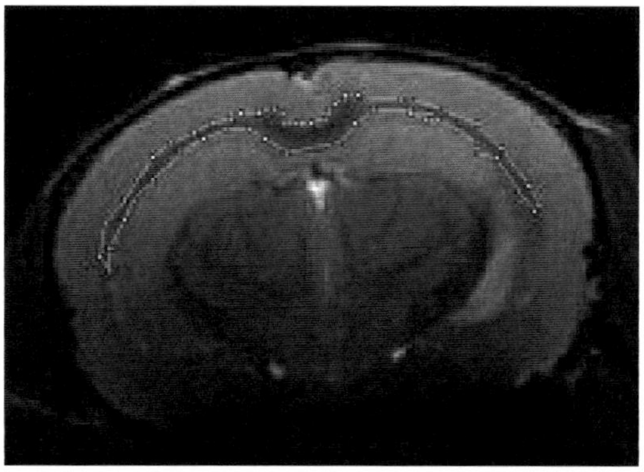

Figure 1. Delineation of the area of the corpus callosum using ImageJ software.

2.7 Statistical analysis
The significant differences between the groups were evaluated using a *t*-test for independent samples.

3 Results
Typical T2-weighted images of mice in the "Control", "Demyelination" and "Demyelination+Ropren" groups are shown on Figure 2. The images demonstrate a noticeable reduction of the corpus callosum in the mouse from the "Demyelination" group and the absence of visible signs of demyelination in the mouse from the control group and the mouse from the group treated with Ropren®.

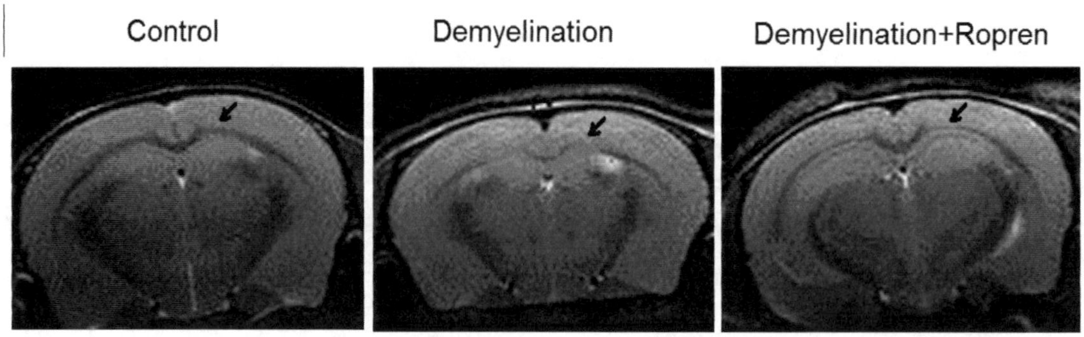

Figure 2. Typical examples of T2-weighted images of mice in the "Control", "Demyelination" and "Demyelination+Ropren" groups. The arrows indicate decreased size of the corpus callosum in the brain of the mouse from the "Demyelination" group and the normal size of the corpus callosum in mice from both the "Control" and "Demyelination+Ropren" groups.

Quantitative analysis showed that animals from the "Demyelination" group had a significantly smaller corpus callosum compared to the "Control" group ($p<0.01$) (Figure 3). CPZ administration reduced the corpus callosum by 19.5% on average. At the same time, mice treated with CPZ and

Ropren® have similar levels of demyelination of the corpus callosum to those seen in the control group, with no significant differences found between the "Control" and "Demyelination+Ropren" groups.

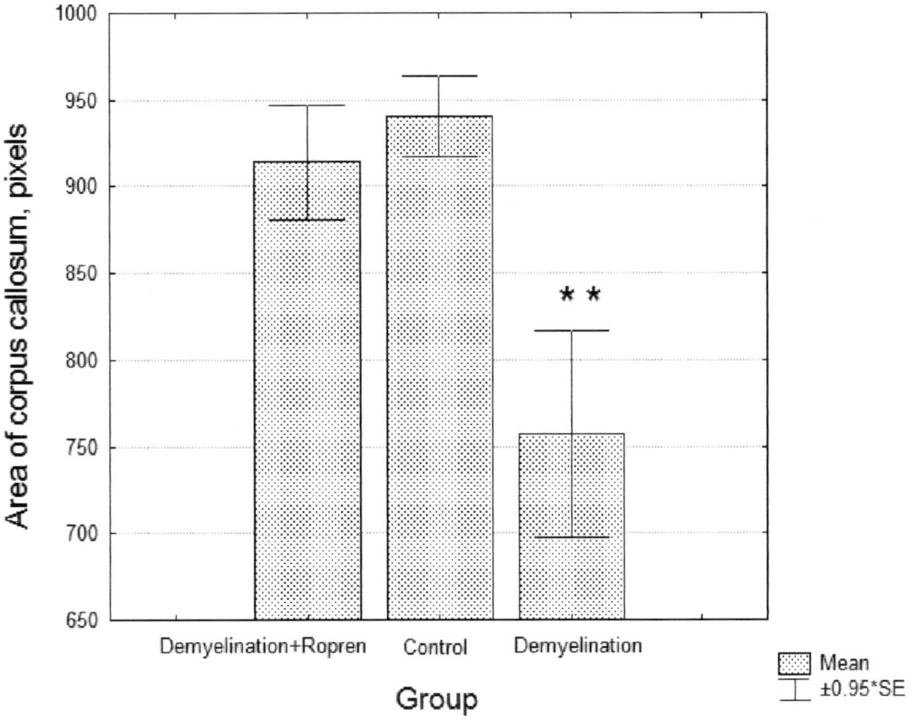

Figure 3. The significant decrease of the size of corpus callosum in cuprizone-treated mice compared to the control group. Mice treated with both Ropren® and cuprizone had similar levels of myelination of the corpus callosum as control animals.
** = p<0.01, *t*-test.

4 Discussion

The significant decrease of the size of the corpus callosum shows successful induction of the CPZ model of demyelination in CD-1 mice. The main finding is that after Ropren® injections, no damage of the corpus callosum was seen after 10 weeks of CPZ administration.

Long-chain polyisoprenoid alcohols, including polyprenols and dolichols, are a unique class of secondary metabolites within the isoprenoid natural product family [6]. Polyprenols and the phosphorylated derivatives comprise a small percentage of the total glycerophospholipid content in cellular membranes of bacteria (~1%) [9] and eukaryotes (~0.1%) [10].

Polyprenyl-phosphates act as oligosaccharide carriers during glycan biosynthesis, which is essential to many conserved cellular processes including N-linked protein glycosylation, C- and O-protein mannosylation, and bacterial cell wall biosynthesis. The production of dolichyl-phosphate is an important regulator of cell differentiation [11]. Inhibition of polyisoprenol biosynthesis resulted in abnormal gastrulation, which correlated with the inability of the cell to produce glycoproteins. Furthermore, addition of exogenous dolichol allowed for normal gastrulation, suggesting that dolichyl-phosphate is a limiting reagent for N-linked glycoprotein biosynthesis and subsequent cellular transformations [12]. The rate of dolichyl-phosphate and glycoprotein synthesis has also been linked to the growth rate of Chinese hamster ovary cells and cell division [13]. Other studies have confirmed that dolichyl-phosphate is a rate limiting substrate in N-linked glycosylation and is thereby a key factor in cellular development [10, 14].

Polyprenols, as well as dolichols, contain polyisoprene units and also exhibit membrane-active properties affecting the viscosity and fluidity of membranes. Apparently, polyprenols could facilitate the membrane transport of proteins or peptides and contribute to nerve impulse transmission by binding glycoproteins with the polyisoprenoid fragment. Moreover, dolichol is involved in the biosynthesis of cholesterol that makes up one-third of the myelin sheath. Sterols and other lipids involved in their biosynthesis (dolichol) are the main agents for proteins passing through the membrane and are involved in the regulation of nerve impulse transduction [15]. Hartley et al. (2012) also suggests that dolichyl-phosphates may influence enzyme and pathway regulation, flippase-mediated glycan translocation across membranes, and macromolecular enzyme complex formation [6].

We suggest that Ropren®, with its composition of approximately 95% polyprenols, may either prevent myelin loss or restore myelin by several pathways. Polyprenols may serve as an exogenous source of isoprenoid chains that form endogenous metabolites, such as dolichol and cholesterol. Another pathway is the biosynthesis of glycoproteins, where phosphorylated derivatives of polyprenols are involved in the process of protein glycosylation and recognition of glycoprotein on the membrane, the recognition of antigens on the membrane, the fluidity and viscosity of membranes, the functioning of ion channels, and nerve impulse conduction, as well as an having an influence on regeneration through cell division, differentiation and development.

Polyprenols can accumulate at relatively high levels within many eukaryotic organisms; biosynthesis of these molecules is a complex, energy-dependent process and it seems unlikely that production of polyprenols would have been conserved in evolution if it did not serve an important role. Further work is necessary to establish the physical effects that polyprenols may exert on complex biological membranes and how polyprenol accumulation affects aging, disease, and intracellular transport processes [6].

5 Conclusions

CPZ treatment caused a significant reduction of corpus callosum associated with demyelination. Mice treated with Ropren® (the plant-derived substance containing 95% polyprenols) and CPZ showed a level of myelination of the corpus callosum similar to the level seen in control animals. The study suggests the positive impact of Ropren® on CPZ-induced demyelination and allow us to consider further testing of this substance as an effective treatment against MS. Future studies will also enable us to determine if Ropren® prevents demyelination or promotes re-myelination.

Acknowledgments

The animal cuprizone model was performed with financial support of Tomsk State University Competitiveness Improvement Program. Ropren® for testing in the cuprizone model was supplied with financial support of Solagran Limited, Biotechnology Company. MRI study was supported by Russian Science Foundation, project № 14-45-00040.

References

[1] Torkildsen O, Brunborg L A, Myhr K-M and Bo L. 2008 *Acta Neurol. Scand.* **117 (188)** 72
[2] Filippi M and Rocca M A 2011 *Radiol.* **259** 659
[3] Gudi V, Gingele S, Skripuletz T and Stangel M 2014 *Front. Cell. Neurosci.* **8** 73.
[4] Soultanov V S, Agishev V G, Monakhova I A, Mokhovikova I A, Kulikov A P, Rosehin V I and Nikitina T V 2010 Gastroenterologia Sankt-Peterburga **4** 12
[5] Surmacz L and Swiezewska E 2011 *Biochem. Biophys. Res. Commun.* **407(4)** 627
[6] Hartley M D and Imperiali B 2012 *Arch. Biochem. Biophys.* **517(2)** 83
[7] Blakmore W F, 1973 J. Neurol. Sci. **20** 63
[8] Fedotova J, Soultanov V, Nikitina T, Roschin V, Ordayn N 2012 *Phytomedicine* **19(5)** 451
[9] Paxinos G and Frankin B J 2001 *The mouse brain in stereotactic coordinates* (San Diego, CA:

Academic Press)
[10] Barreteau H, Magnet S, El Ghachi M, Touze T, Arthur M, Mengin-Lecreulx D, Blanot D, Chromatogr J and Analyt B. 2009 *Technol. Biomed. Life Sci.* **877** 213
[11] Rosenwald A G, Stoll J and Krag S S. 1990 *J. Biol. Chem.* **265** 14544
[12] Lucas J J and Levin E 1977 *J. Biol. Chem.* **252** 4330
[13] Rossignol D P, Lennarz W J and Waechter CJ. 1981 *J. Biol. Chem.* **256** 10538
[14] Kabakoff B D, Doyle J W and Kandutsch A A 1990 *Arch. Biochem. Biophys.* **276** 382
[15] Spiro M J and Spiro R G 1986 *J. Biol. Chem.* **261** 14725
[16] Schroeder F, Atshaves B P, McIntosh A L, Gallegos A M, Storey S M, Parr R D, Jefferson J R, Ball J M, and Kier A B 2007 *Biochem.Biophys.Acta*, **1771 (6)**, 700

Journal of Physics: Conference Series **677** (2016) 012008 doi:10.1088/1742-6596/677/1/012008

MR histology of advanced atherosclerotic lesions of ApoE-knockout mice

A Naumova[1,2], V Yarnykh [1,2], M Ferguson[1], M Rosenfeld[3], C Yuan[4]

[1] Vascular Imaging Lab, Department of Radiology, University of Washington, Seattle, WA, USA
[2] Neurobiology Lab, Research Institute of Biology and Biophysics, Tomsk State University, Tomsk, Russian Federation
[3] Department of Pathobiology, University of Washington, Seattle, USA
[4] Departments of Radiology and Bioengineering, Center for Cardiovascular Biology, Institute for Stem Cell and Regenerative Medicine, University of Washington, Seattle, WA, USA

E-mail: nav@uw.edu

Abstract
The purposes of this study were to examine the feasibility of determining the composition of advanced atherosclerotic plaques in fixed ApoE-knockout mice and to develop a time-efficient microimaging protocol for MR histological imaging on mice. Five formalin-fixed transgenic ApoE-knockout mice were imaged at the 9.4T Bruker BioSpec MR scanner using 3D spoiled gradient-echo sequence with an isotropic field of view of 24 mm^3; TR 20.8 ms; TE 2.6 ms; flip angle 20°, resulted voxel size 47 x 63 x 94 μm^3. MRI examination has shown that advanced atherosclerotic lesions of aorta, innominate and carotid arteries in ApoE-knockout mice are characterized by high calcification and presence of the large fibrofatty nodules. MRI quantification of atherosclerotic lesion components corresponded to histological assessment of plaque composition with a correlation coefficient of 0.98.

1. Introduction

Transgenic mice are widely used as animal models for the study of atherosclerosis. Apolipoprotein E–knockout (ApoE-/-) mice produced by gene-targeting technologies spontaneously develop atherosclerotic lesions similar in morphology to those observed in humans [1, 2]. The major challenges for histological MR imaging of atherosclerotic lesions in mice are the very high-resolution requirement, the need to image long vessel segments, and prohibitively long scan time. Several attempts were made to develop MRI protocols for imaging of atherosclerotic plaques in ApoE-/- mice [3-6]. None of these studies have demonstrated the ability of histological MR to characterize the clinically significant advanced atherosclerotic lesions. The purposes of this study were to examine the feasibility of determining the composition of advanced atherosclerotic plaques in fixed ApoE-/- mice and to develop a time-efficient microimaging protocol for MR histological imaging on mice.

2. Methods

2.1. Specimen preparation:
Five transgenic ApoE-/- mice (75 weeks old) were perfusion-fixed at physiologic pressure via the left ventricle (24-g ¾-in Becton Dickinson Angiocath intravenous catheter). The animals were initially perfused with 10mL PBS for 20 seconds followed by a 4-minute perfusion of 10% buffered formalin (Fisher Scientific, Fairlawn, NJ) along with Omniscan (gadodiamide; Amersham Health Inc, NJ), a conventional clinical contrast agent at a volume ratio of 20:1.

2.2. MRI:
Experiments were performed using a 9.4T Bruker BioSpec MR scanner. 3D spoiled gradient-echo sequence was used with an isotropic FOV of 24 mm^3; 3D matrix of 512x384x256; TR=20.8ms; TE=2.6ms; flip angle=20°. All MR images had a voxel size of 47x63x94μm^3. The total scan time was 1 hour 8 min. These scan parameters allowed coverage of major vasculature spanning from the innominate artery bifurcation to the upper margin of the liver.

2.3. Histology:
After imaging, the mouse was decalcified in 10% formic acid and then routinely processed en block through graded alcohols, xylene and paraffin. The specimen was embedded in paraffin and 10 micron serial section was collected every 50 microns throughout the whole mouse thorax and stained with Hematoxylin-Eosin. Images obtained with magnetic resonance technique were compared to matched corresponding histology. Anatomic structures such as the aorta, common carotid arteries, spinal cord, myocardium, aortic valve were all used as external fiducial references for alignment of the MR images with the corresponding histology sections. Tissue components in atherosclerotic lesions were identified by histology and then their MR signal properties were characterized.

3. Results

Advanced atherosclerotic lesions were found by MR examination in the aorta, aortic arch, innominate and carotid arteries of the mice. Atherosclerotic plaque characteristics, such as plaque area, volume, thickness, and major plaque components, such as necrotic core, calcification, and fibrous tissue were determined. Morphologically, the lesions found in the MR images corresponded closely to the histopathological cross-sections taken from the same location (figure 1). Preliminary results of a pilot study quantifying atherosclerotic lesion components produced a correlation coefficient of 0.98. MR histology, in agreement with prior studies [7, 8], has shown that advanced atherosclerotic lesions of aorta, innominate and carotid arteries in ApoE-/- mice are characterized by high calcification and presence of the large fibrofatty nodules.

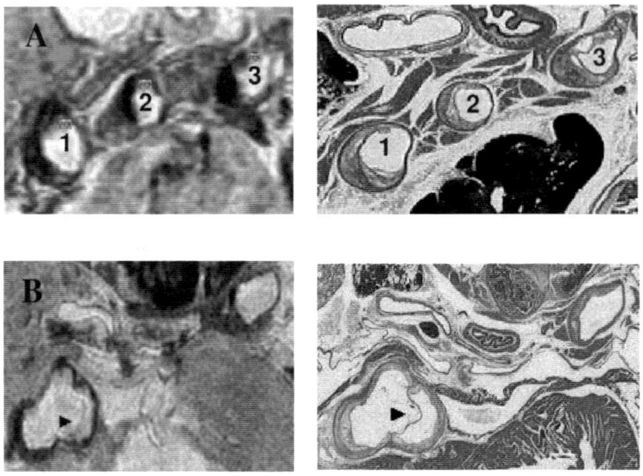

Figure 1. Cross-sectional 1H MR images and corresponding histological slices through the mouse thorax at the (A) level of ascending from the aortic arch arteries (1 - innominate artery; 2 - left common carotid artery; 3 - left subclavian artery) and (B) at the level of the aortic root showing a clearly defined aortic valve (arrowhead).

4. Conclusion

This study has shown an ability of advanced atherosclerotic lesions characterization by MR histology on the murine model of atherosclerosis. High-field MRI allows non-invasively determination of atherosclerotic plaque composition and precise quantification of the major plaque components in the different vascular beds. A time efficient MRI protocol, based on a single contrast weighting, was developed using fixed ApoE-/- mice and can be used for in vivo studies.

Acknowledgements

NIH T32 BCTG and Amgen Inc for MRI scans and histological analysis, Russian Science Foundation (project № 14-45-00040) for data analysis, and Tomsk State University Competitiveness Improvement Program for manuscript preparation. We thank Jerry Ricks for performing of ApoE-/- mice perfusion/fixation.

References:

[1] Plump A S, Smith J D, Hayek T, et al 1992 *Cell* **71** 343–353
[2] Zhang S H, Reddick R L, Piedrahita J A, Maeda N 1992 *Science* **258** 468–471
[3] Chaabane L, Soulas E C, Contard F, et al 2003 *Invest Radiol* **38(8)** 532-538
[4] Wiesmann F, Szimtenings M, Frydrychowicz A, et al 2003 *Magn Reson Med* **50(1)** 69-74
[5] Itskovich V V, Choudhury R P, Aguinaldo J G, et al 2003 *Magn Reson Med* **49(2)** 381-385
[6] Schneider J E, McAteer M A, Tyler D J, et al 2004 *J Magn Reson Imaging* **20(6)** 981-989
[7] Rattazzi M, Bennett B J, Bea F, et al 2005 *Arterioscler Thromb Vasc Biol* **25(7)** 1420-1425
[8] Rosenfeld M E, Polinsky P, Virmani R, et al 2000 *Arterioscler Thromb Vasc* Biol **20(12)** 2587-2592

The use of innovative gadolinium-based contrast agent for MR-diagnosis of cancer in the experiment

V Chernov[1,2], A Medvedeva[1,2], I Sinilkin[1,2], R Zelchan[1,2], E Grigorev[1], I Frolova[1], I Nam[2]

[1]Cancer Research Institute, Siberian Branch of the Russian Academy of Medical Sciences, Tomsk, Russian Federation
[2]National Research Tomsk Polytechnic University, Tomsk, Russian Federation

E-mail: r.zelchan@yandex.ru

Abstract. The present study of the functional suitability and specific activity of the contrast agent gadolinium-based for magnetic resonance imaging demonstrated that the investigated contrast agent intensively accumulates in organs and anatomical structures of the experimental animals. In the model of tumor lesions in animals, study have shown that investigational contrast agent accumulates in the tumor tissue and retained there in for a long enough time.

1. Introduction

Modern technology of magnetic resonance imaging (MRI) is closely associated with the use of magnetic resonance contrast agents (MRCA), which not only increase the sensitivity and specificity of the MRI diagnosis, but also are an integral part of the diagnostic procedure. In European countries, the MRCA are used in 70-80% of cases, in Russia this indicator is 9%, primarily due to economic reasons and the lack of domestic contrast agents. The study of tumors without MRCA has low diagnostic information, so according to the WHO, at the MRI without contrast misdiagnosis rate may reach 40%.The most frequently used in the practice of MRI paramagnetic metal is gadolinium (Gd + 3), which has seven unpaired electrons [1-6].

Solving the problem of creating a domestic gadolinium-based agent for magnetic resonance diagnostics will provide national health care effective MRCA cheaper foreign contrasts.

Therefore, the aim of the study was to assess the possibility of experimental use of a domestic contrast agent on the basis of a gadolinium complex "Pentagascan" for magnetic resonance diagnosis of malignant tumors.

2. Materials and methods

The MRSA ("Pentagascan" LLC "MedContrastSyntes") was studied. Agent is disodium gadolinium diethylenetriamine-N, N, N', N", N", - -pentaacetic acid as a solution for infusion (Table. 1).

Table 1. Information about the "Pentagascan".

Indicator	Regulated values
Appearance	Clear yellowish liquid
Authenticity	Matches
pH	6.8–7.4
The color of the solution	Not more Y5
Mechanical impurities	Absence
The relative density	1.120–1.240
The content of total gadolinium, mg/cm3	75–90
The content of the disodium salt of the complex gadolinium DTPA, mg/cm3	282–338
Excess DTPA, mg / cm3, not more than	10
Free Gd+ of the total, %, not more than	0.3
Heavy metals,%, not more than	0.001
Bacterial endotoxins, not more than	26 EU/см3
Sterility	Sterile

The first phase of the study was a comparative assessment of the suitability of the "Pentagascan" as MR contrast agent in intact animals. Comparison drug was "Magnevist". For this purpose 60 intact male rats white line "Wistar" weighing 300-350 g were studied. First, MRI was performed in all animals in native mode, without the introduction of paramagnetic agents. Then the animals were divided into two groups depending on the contrast agent. The study was performed in intact rats at 5 and 15 min after administration of the contrast agent into the femoral vein in a dose of 0.2 ml.

The second phase of the study was to evaluate the efficacy of "Pentagascan" in animals with experimental tumors. As a biological tumor model used malignancies transplanted mice - Lewis lung carcinoma. A solid version of Lewis lung carcinoma was used (Bank of cell lines RCRC. NN Blokhin, Moscow), supported by in vivo in mice lines C57B1 / 6j by intramuscular transplantation. To obtain tumor cells, mice were sacrificed by cervical dislocation. Tumor tissues were removed using sterile instruments, homogenized and counted in a cell concentrations Goryaev chamber. Cells were transplanted intramuscularly to the back the right paw at a concentration of 1-3 million cells per mouse.

All invasive procedures with animals were carried out with the use of inhaled or medication anesthesia. During this phase of the study 60 mice of the C57B1/6j were included in the experiment, at the time of the survey the mass of animals was 30-35, the average volume of the tumor node, measured instrumentally, was 2.3 ± 0.7 cm3.

All the animals were kept on a standard diet of the vivarium with free access to water, in accordance with the rules of the European Convention for the Protection of Vertebrate Animals used for experimental and other scientific purposes.

Studies of mice with transplanted tumor were performed before and 1, 5 and 9 min after injection of contrast "Pentagascan" in a volume of 0.1 ml in the infraorbital sinus. In order to ensure complete immobility throughout the study experimental animals were anesthetized with an intramuscular infusion of 5% solution of ketamine.

Animal disposal on a pallet in a natural position. Centre tray with the animal was placed in a tunnel of the magnet isocenter, it used a 16-channel head coil HIGH-Head Matrix. The images were obtained with the following parameters Scan: T1-SE sequence - TR = 550 ms, TE = 17 ms, FoV - 181 × 270 mm, matrix - 292 × 512, slice thickness - 5 mm step - 1.5 mm; sequence of T1-TSE fatsat - TR = 677 ms, TE = 11 ms, FOV - 135 × 270 mm, matrix - 512 × 192, slice thickness - 3 mm, pitch - 0.3 mm.

Contrasting effect of the agent assessed visually and semi-quantitatively.

Statistical processing of the obtained data was performed using the software package STATISTICA. To estimate the differences of quantitative traits between groups a nonparametric Mann-Whitney test was used. Statistical significance of the dynamics of change in quantitative indicators was estimated using paired Wilcoxon test. In all procedures of the statistical analysis the significance level p was taken to be 0.05. Results are presented as M ± σ, where M is the arithmetic mean, σ - standard deviation.

3. Results and discussion

By visual evaluation of intact animal images after the administration "Pentagascan" and the post-injection "Magnevist" enhancement MR signal on post contrast scans in T1 images was marked. Contrast agent accumulated in the kidney cortex and vessels. On the delayed scans visualized homogeneous accumulation of contrasts in the kidney, liver and spleen. Also accumulation of paramagnetic in the bladder was determined (Fig. 1, 2).

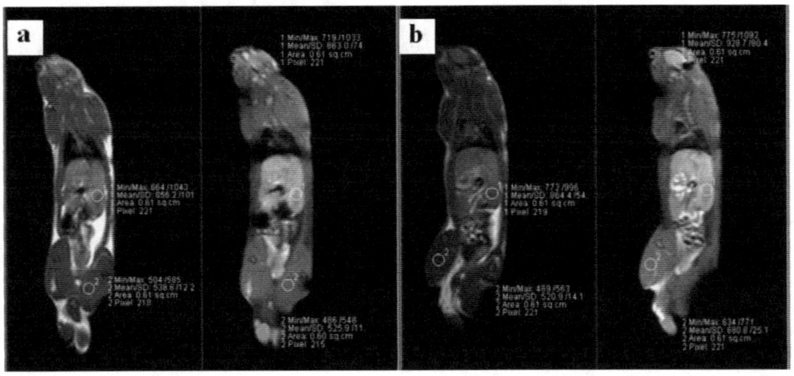

Figure. 1 Distribution of study drug ("Pentagaskan") in the body intact animals. After contrast injection modes T1-SE (a) and T1-TSEfatsat (b), there is its accumulation in renal parenchyma, liver, bladder.

Journal of Physics: Conference Series **677** (2016) 012009 doi:10.1088/1742-6596/677/1/012009

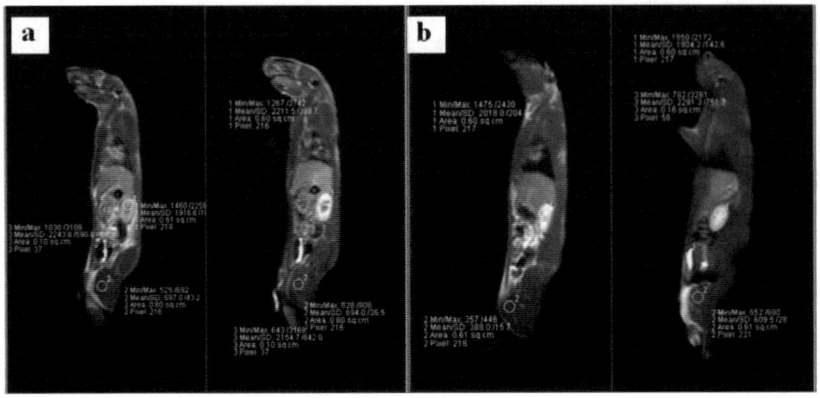

Figure. 2 Distribution of means of comparison ("Magnevist") in the body of intact animals. After contrast injection modes T1-SE (a) and T1-TSEfatsat (b), there is its accumulation in renal parenchyma, liver, bladder

The intensity of MR signal was measured built-in software module, by allocating ROI (region of interest) in the same area as a circle covering the cortex and medulla of the kidney. Considering the thickness of the slice, a volume of tissue in the region of interest was of 0.60-0.61 cm2.

As a reference area used muscles (hip area). The MR signal intensity in the respective portions of the body are shown in Table. 2.

Table 2. The intensity of the MR signal.

	The value of the measured intensity in the selected areas of interest (value / standard deviation)				
	Prior to the introduction		After contrast administration		Contrasted urine in the bladder
	kidney	muscle	kidney	muscle	
"Pentagaskan" T1-SE	825.7/14.8	530.3/5.1	1902.3/28.7	565.8/10.3	2192.9/9.5
"Magnevist" T1-SE	864.4/54	520.9/14	2018.0/204	388.0/15	2142.0/446
"Pentagaskan" T1-TSE fatsat	879.8/40.1	580.6/83.1	2111.1/181	665.5/54.3	2188.2/34.4
"Magnevist" T1-TSE fatsat	928.7/80	680.8/25	1904.2/143	609.5/28	2291.3/751

The specific activity research study of "Pentagaskan" in intact animals indicate that contrasting ability of this MRCA does not differ from “Magnevist" the results of visual and semiquantitative evaluation.

In order to determine the optimal time intervals for the tumor study after intravenous contrast enhancement using a “Pentagascan", it was carried out a series of experiments. The intensity and character of accumulation of contrast agent in tumor tissue at various time intervals: at 1, 5 and 9 min after intravenous contrast injection was visually evaluated. Time intervals are selected based on the intensity of metabolic processes in experimental animals.

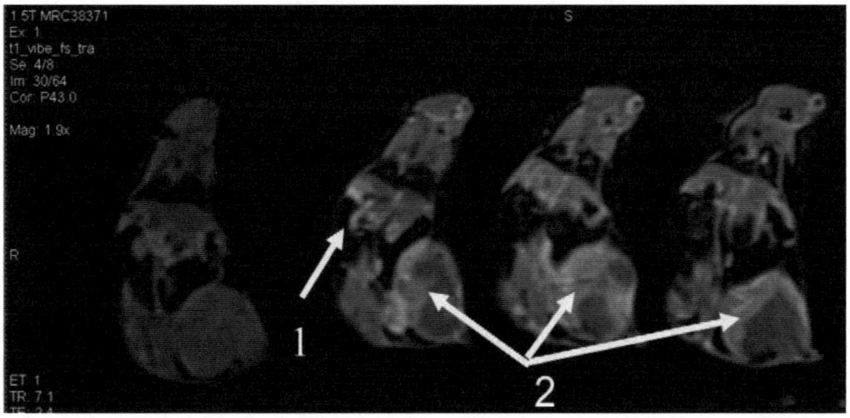

Figure 3. MR tomography animals in the mode T1-VIBE FS (left - right): mouse without contrast enhancement; through 1; 5 and 9 minutes after the administration of contrast. It noted more intense accumulation of paramagnetic material renal parenchyma in the early phase of the scan (1). Nonuniform (due to evidence of necrosis in the central regions of the tumor), tumor accumulation of contrast (2)

It was noted that in the early phase "Pentagascan" rapidly accumulate in the renal parenchyma. The tumor tissue uptake of the MRCA has been observed in the early phase of the scan and stored in it for a long time. Intensity of drug accumulation in the tumor remained at the same level in different phases of the study (Fig. 3).

For confirmation of contrast enhancement of "Pentagaskan" quantitative parameters of the MR signal in the native phase of the study and after the intravenous administration of MRCA were calculated. The level MR signal in the tumor and in the kidney before and after intravenous contrast was measured. Mean values of MR signal intensity in tumor node and in the projection of the kidney are presented in Table 3.

Table 3. MR signal intensity in experimental animal studies in native phase and after contrast enhancement.

The average values of the measured intensity in the selected regions of interest (value/standard deviation)			
Kidney		Tumor	
Native phase	Contrast enhancement	Native phase	Contrast enhancement
931.0/23.2	1929.0/16.9	561.9/11.2	1039.7/8.9

4. Conclusion

"Pentagascan" intensively accumulates in organs depleting drug that reliably fixed with any standard T1-weighted scans;

"Pentagaskan" intensively uptakes and stores for a long time in the pathological tumor tissue, which allows to perform quality MR scans of tumors.

References

[1] Brasch R C, Weinmann H-J and Wesbey G E 1984 *AJR.* Vol. 142 (3) p 625–630
[2] Carr D H, Brown J, Bydder G M, Weinmann H J, Speck U, Thomas D J and Young I R 1984 *Lancet*. Vol. 1 (8375) pp 484–486
[3] Laniado M, Weinmann H J, Schörner W, Felix R and Speck U 1984 *Physiol. Chem. Phys. Med. NMR.* Vol. 16 (2) pp 157–165
[4] Weinmann H J, Brasch R C, Press W R and Wesbey G E *AJR.* Vol. 142 (3) pp 619–624

[5] Weinmann H J, Ebert W, Misselwitz B and Schmitt-Willich H 2003 *Eur. J. Radiol.* Vol. 46 (1) pp 33–44
[6] Weinmann H J, Laniado M and Mützel W. 1984 *Physiol. Chem. Phys. Med. NMR*. Vol. 16 (2) pp 167–172

3D DCE-MRA of pedal arteries in patients with diabetes mellitus

M Zamyshevskaya[1], V Zavadovskaya[1], M Zorkaltsev[1], V Udodov[1], E Grigorev[1]

[1] Department of Radiology and Radiotherapy, Siberian State Medical University, Tomsk, Russian Federation

E-mail: zamyshevskayamari@mail.ru

Abstract. Purpose was identification and evaluation of pedal vascularization in diabetic patients of using contrast MR-angiography (3D DCE-MRA). 23 diabetic feet of 23 patients (15 male, 8 female; mean age 56±14.6) underwent 3D DCE-MRA (Gadobutrol 15ml) at 1.5 T. Imaging analysis included blood-flow's speed, vascular architectonic's condition and character of contrast's accumulation. Osteomyelitis was verified by surgery in 15 cases. All patients were divided in 3 groups: neuropathic, neuroischemic, ischemic forms of diabetic foot. First-pass MRA detected significant delay of contrast's arrival in ischemic group. There were no significant differences between the values of neuropathic and neuroischemic forms of diabetic foot. Pedal vessels in patients were absent. Contrast MRA revealed three types of contrast distribution in soft tissues: uniform, local increase and local absence. Osteomyelitis was associated with diffuse enhanced contrast accumulation in all cases. In summary, MRI blood vessel imaging is a promising and valuable method for examining peripheral arterial changes in diabetic foot and might be useful for treatment planning in different forms of diabetic foot.

1. Introduction

Diabetic foot syndrome (DFS) is one of the most common and serious complications of diabetes mellitus [1]. This syndrome includes arterial and microvasculature involvement, neurotrophic disorders and is accompanied by bone and soft tissue structural changes. Infection and the subsequent development of osteomyelitis are often associated with high risk of amputation. High mortality in patients with DFS after amputation is well known [2].

The main conditions for reducing the amputations and mortality are early detection and subsequent treatment of osteomyelitis, which is impossible without blood-flow correction [3]. Radiological methods, including selective digital subtraction angiography (DSA) as the "gold standard", widely used in evaluation of the main and peripheral pedal blood vessels [4]. However, despite the fact that DSA is a standardized method, it has some limitations, such as invasiveness, relatively high radiation dose, and the possibility of contrast-induced nephropathy in such patients.

Magnetic resonance angiography (MRA) is an alternate method, characterized by high spatial resolution and absence of nephrotoxic effects of the contrast media. Many publications show the advantages of using MRA for DSA patients [5]. However, majority of publications focuses on vascular permeability; MRA data on DFS patients with complicated purulent necrotic process are limited.

The pupose of the study was to characterize pedal arteries (vascular architectonics, contrast arrival time and character of contrast accumulation) in patients with complicated diabetic foot using contrast-enhanced MRA.

2. Materials and methods

Our study was full-design. 23 patients (15 men, 8 women, mean age 56 ± 14.6 years) with neuropathic (n = 9; 39.0%), mixed (n = 10; 43.5%), ischemic (n = 4; 17.5%) DFS forms with suspected osteomyelitis (OM) underwent 3D DCE-MRA (Gadobutrol 15ml) at 1.5 T MR imaging unit (Siemens MAGNETOM Essenza, Germany). The diagnosis of diabetes mellitus was established according to WHO criteria, 1999. All patients complained of pain in the affected foot. An objective study identified the following symptoms: swelling of the soft tissues (100%), hyperaemia (82.6%), ulcer (43.5%) or fistula (26.1%), local temperature increase (69.6%), the weakening of pedal vessels pulsation (34.8%). All patients had either normal or only slightly reduced renal function with estimated glomerular filtration rates (GFR)>30 ml/kg body weight. Osteomyelitis was verified according to operations in 15 cases.

Magnetic resonance imaging of ankle and foot was carried out using MR-tomography 1.5 T clinical scanner in three standard planes - sagittal, coronal and transversal, slice thickness of 3-5 mm, by using T1-WI and T2-WI, FSat sequences. After receiving the native image MRA with contrast agent bolus Gadovist (Bayer HealthCare Pharmaceuticals) in 15 ml was performed. Angiographic images were recorded within 180-300 seconds since the contrast administration by taking cyclically repeating series of 17-second scans, which were subsequently processed using subtraction and building MIP-reconstructions. It was performed delayed postcontrast MR - study of feet using sequences T1-WI for comparison angiography and perfusion.

3. Results

The study included three stages. In the first stage the contrast arrival time in pedal vessels was assessed, which was averaged 21 ± 11.4 sec according to a study in the total group. At the same time, was revealed significant difference of contrast arrival time in DFS ischemic form, which was 53 ± 33.15 seconds. Approximately the same values were obtained for neuropathic form - 20 ± 12,5 sec and mixed form - 19 ± 9,3 sec.

Next, using as reference data Martin R. Prince (2001) [6], all patients were divided into 3 groups: with timely (n = 11, 48%), accelerated (n = 8, 35%) and delayed (n = 4, 17%) contrast arrival in pedal vessels. Accelerated contrast arrival observed equally in neuropathic (n = 4, 50%) and mixed (n = 4, 50%) forms DFS [Fig. 1a,b].

Delayed contrast arrival was determined mainly in ischemic form of DFS (n=3, 75%) and only in 1 case (25%) it was identified in the neuropathic form of DFS.

In the second stage of the study was evaluated pedal vascular architectonic's condition. In this study the normal vascular pedal vascular architectonics was established in 12 (52.2%) cases. It was revealed visualization absence of aa. digitales dorsales in 12 (52.1%) cases, mainly in patients with ischemic (100%) and mixed (40.0%) DFS forms [Fig. 2].

In rare cases, there was no visualization of arcus plantaris (n = 1, 4.3% mixed form), a. arcuata (n = 1, 4.3%; ischemic form), a. metatarsea dorsales of 2nd and 3rd fingers (n = 1, 4.3% mixed form).

Another feature of the circulation in the studied group of patients was an early visualization of veins [Fig. 1b]. This picture has been diagnosed in 8 patients (34.8%) with neuropathic (n=6, 66.7%) and mixed forms of DFS (n=2, 20.0%).

In the third stage was studied character of pedal soft tissue contrast accumulation reflecting the state of the microcirculation. All patients were divided into 3 groups: with uniform (n = 2, 8.7%), local increase (n = 20, 87.0%), local absence (n = 1, 4.3%) of contrast accumulation. Local increase of contrast accumulation visualized in two variants: until distinct pedal vascular imaging (n = 3, 15.0%) and simultaneously with the appearance of pedal vascular architectonic (n = 17, 85.0%).

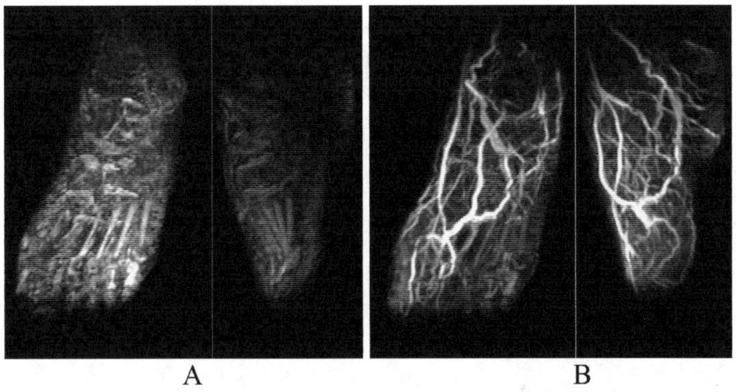

Figure 1. Patient K., 54 years old, diabetes type 2, mixed form of DFS. Objective: foot swelling, hyperemic, with ulcers on the medial surface of the 1st finger of the right foot. MRA: **A.** Early opacification diffuse soft tissue to distinct visualization of the main arteries of the foot (0-17 sec). **B.** Early visualization of the venous outflow - an accelerated arteriovenous transport (18-34 sec).

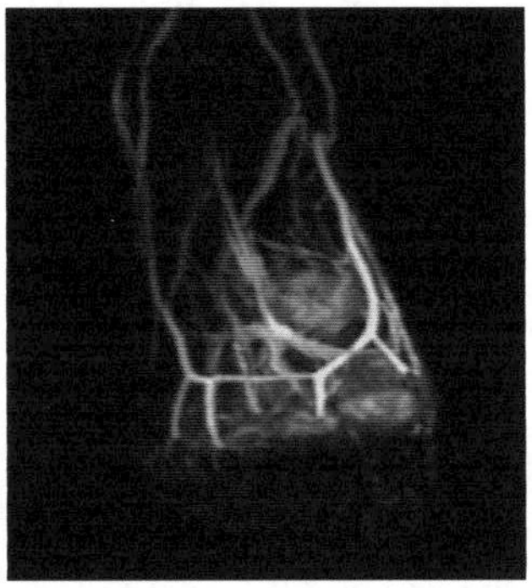

Figure 2. Patient K. 29 years, type 1 diabetes, a mixed form of VTS, suspected osteomyelitis. MRA: absence of aa. digitales visualization.

Uniform soft tissue contrast accumulation was observed equally in ischemic and neuropathic forms of DFS (n = 1, 4.3% respectively). Local increased accumulation was detected only in patients with neuropathic (n = 2, 8.6%) and mixed (n = 1, 4.3%) forms. Increased contrast local accumulation simultaneous with the vessels visualization principally was detected in the mixed (n = 8, 34.8%) and neuropathic (n = 6, 26.1%) DFS forms, ischemic form was characterized by a smaller number of cases (n = 3, 13.0%) [Fig. 3].

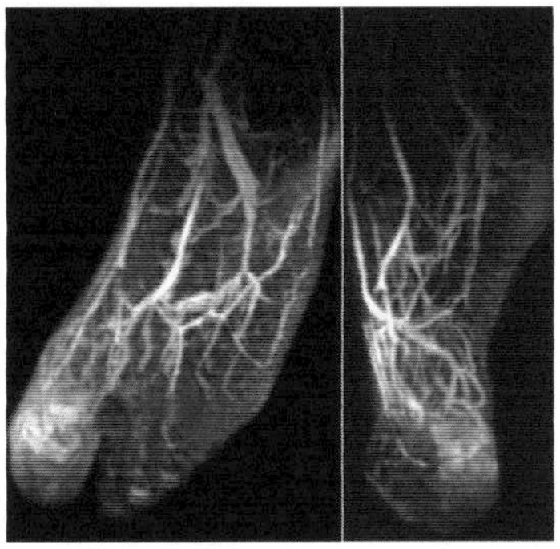

Figure 3. Patient L., 27 years old, diabetes type 1, neuropathic form of DFS. Objective: hyperemia, swelling, ulcer with purulent contents of the medial surface of the 1st left toe. MRA: Local long-term increased contrast accumulation in the soft tissues of the 1st left toe without clear visualization of vascular architectonics.

Journal of Physics: Conference Series **677** (2016) 012010 doi:10.1088/1742-6596/677/1/012010

Absence of contrast accumulation was detected in the case of mixed form (n=1, 4.3%) in a patient with gangrene of the 1st finger of the left foot [Fig. 4a].

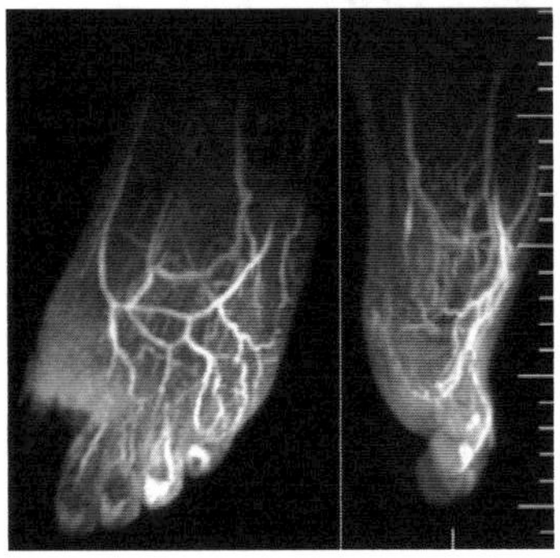

Figure 4. Patient E., 54 years old, diabetes type 2, a mixed form of DFS. Objective: gangreen of the 1st left toe. MRA: throughout the study, up to five minutes absence of contrast accumulation in the 1st left toe.

4. Discussion

Complications of diabetic foot, neuropathy caused by ischemia and joining infections are a common cause of amputations of the lower limbs, which leads to severe disability and makes a significant contribution to mortality. Early diagnosis enables timely surgical or radiological intervention and revascularization in 80% of cases [7]. 3D contrast-enhanced MR-angiography is rapidly gaining acceptance as a versatile noninvasive alternative to conventional angiography and has been reported as an accurate technique for analyzing the foot arteries in patients with diabetes [8].

At the first stage of study delayed contrast arrival time was revealed principally in ischemic form of DFS and conversely, accelerated blood-flow - in mixed and neuropathic forms of DFS, which corresponds to the pathogenesis.

When comparing the average contrast arrival time depending on the form of DFS statistically significant difference between the rates for ischemic and neuropathic ($p < 0{,}005$) and between the ischemic and mixed ($p < 0{,}031$) DFS forms were identified. At the same time there no statistically significant differences between the values of the average contrast arrival time in neuropathic and mixed forms of DFS were found.

In the second phase in the assessment of vascular architectonics condition it was revealed the absence of several pedal vessels branches in ischemic and mixed forms of DFS, which corresponds to the literature data. In addition, mostly for neuropathic and, to a lesser extent, in mixed forms was determined early visualization of the pedal veins due to arteriovenous shunting. Thus in 5 (62.5%) of 8 cases arteriovenous shunt, this picture has been associated with purulent inflammation.

In the third stage - evaluation the character of pedal soft tissue contrast accumulation - particular attention was paid for long-term diffuse enhanced contrast accumulation in neuropathic and mixed forms of DFS, preferably (n = 16, 80.0%) was accompanied by a purulent inflammation. However, a local increase of contrast accumulation is not specific feature for the presence of purulent inflammatory process and occurs in other clinical situations, including the Charcot foot (n = 4, 20%). Particularly interesting are the cases of extravasates before visualization of pedal vascular architectonics (n = 3, 15%), due to, in our opinion, significantly increased vascular permeability [Fig. 5a, b].

5. Conclusions

In assessing the 3D DCE-MRA patterns in patients with diabetic foot syndrome the following changes in pedal vascularization were found: absence of several pedal vessels visualization in patients with ischemic and mixed forms of diabetic foot syndrome; early visualization of the veins due to arteriovenous shunting; the combination of delayed arrival time of contrast agent with the presence of purulent inflammation in diabetic foot syndrome in some cases; the presence of long-term diffuse enhanced contrast accumulation in soft tissues associated with osteomyelitis.

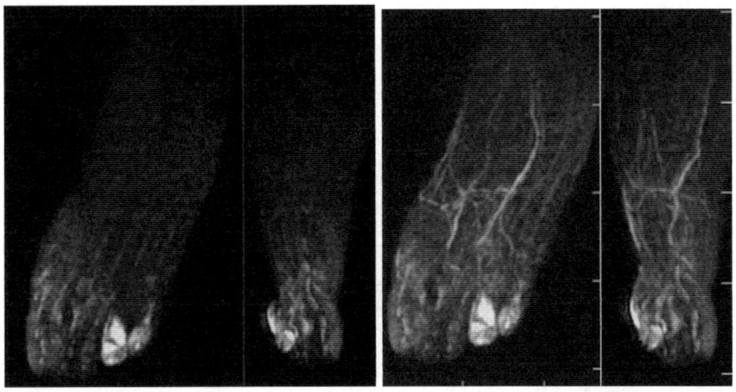

Figure 5. Patient P., 53 years old, diabetes type 2, ischemic form of DFS. Objective: hyperemia, swelling, ulcer with purulent contents of the medial surface of the 4th left finger. MRA: **A.** Local increase contrast accumulation of in the 4 left toe before visualization of main pedal vessels (18-34 seconds). **B.** Local increase contrast accumulation of in the 4 left toe stored for a long time (1.28-1.44 sec).

In summary, MRI blood vessel imaging is a promising and valuable method for examining peripheral arterial changes in diabetic foot; it might be useful for treatment planning in different forms of diabetic foot.

References

[1] Ametov AS Diabetes mellitus type 2 2014 *Problems and Solution Moscow: GEOTAR-Media* **1032** (in Russian)

[2] Malhotra R, Chan CS, Nather A 2014 Osteomyelitis in the diabetic foot *Diabet Foot Ankle* **30** 5

[3] Bargellini I, Piaggesi A, Cicorelli A, et al 2013 Predictive value of angiographic scores for the integrated management of the ischemic diabetic foot *J Vasc Surg* **57(5)** 1204-12

[4] Manzi M, Cester G, Palena LM, et al 2011 Vascular imaging of the foot: the first step toward endovascular recanalization *Radiographics* **31(6)** 1623-36

[5] Röhrl B, Kunz RP, Oberholzer K, et al 2009 Gadofosveset - enhanced MR angiography of the pedal arteries in patients with diabetes mellitus and comparison with selective intraarterial DSA *Eur Radiol* **19(12)** 2993-3001

[6] Prince MR, Wang Y, Watts R, et al 2001 Contrast travel times measured on 2D Projection MRA in patients with Peripheral Vascular Disease *Proc. Int Soc Mag Reson Med* **9** 47

[7] Ranachowska C, Lass P, Korzon-Burakowska A, Dobosz M 2010 Diagnostic imaging of the diabetic foot. *Nucl Med Rev Cent East Eur* **13(1)** 18-22

[8] Li J, Zhao JG, Li MH 2011 Lower limb vascular disease in diabetic patients: a study with calf compression contrast-enhanced magnetic resonance angiography at 3.0 Tesla *Acad Radiol* **18(6)** 755-63

Magnetic Resonance Imaging (MRI) with retrograde intra-lumen contrast enhancement of the rectum in diagnostics of rectovaginal fistulas after combination therapy of rectal cancer. Experience of application

A Usova[1,2], I Frolova [1], S Afanas'ev[1], A Tarasova[1], S Molchanov[1]

[1]Cancer Research Institute, Siberian Branch of the Russian Academy of Medical Sciences, Tomsk, Russian Federation

[2] Neurobiology Lab, Research Institute of Biology and Biophysics, Tomsk State University, Tomsk, Russian Federation

E-mail: afina.tsk@gmail.com

Abstract. Experiment of use of MRI in diagnostics of rectovaginal fistulas after combination therapy of rectal cancer is shown on clinical examples. We used retrograde contrasting of a rectum with 150ml ultrasonic gel to make MRI more informative in case of low diagnostic efficiency of ultrasound, colonoscopy and gynecological examination.

1. Introduction

Rectal cancer accounts for 2-10% of all cancer cases and for 25-30% of all gastrointestinal tumors. Surgery remains the treatment of choice for rectal cancer. The combination rectal cancer therapy includes radiation/chemoradiation therapy which increases the risk of surgical complications [1]. Failure of entero-entero anastomosis sutures after resection of the rectum is an important concern. The rate of this complication reaches 12.6-18.6% [2, 3, 4]. Rectovaginal fistulas after resection of the rectum occur in 3–5.2% of cases [5, 6] and need to be corrected surgically. Rectal and gynecological examinations are not always informative. Ultrasound studies in circumstances of postoperative and postradiation changes are not always effective for a detailed evaluation of anastomosis. Videocolonoscopy identifies anastomosis defects almost in 100% of cases; however, it fails to evaluate surrounding tissues. MRI is a method of choice in diagnosis of pathologies in the pelvis [7,8]. In case of rectal cancer, MRI is a mandatory standard of examination prior to and after the therapy [9,10]. After the completion of the therapy, MRI is used for diagnostics of recurrences and evaluation of postoperative changes [11,12,13,14]. Despite its high resolution, MRI sometimes fails to detect small defects of the wall in the postoperative and postradiation period. Only a few studies analyze the effectiveness of MRI for postoperative complications [15,16]. At the present time, there is no effective method for the diagnostics of rectovaginal fistulas and surrounding tissues.

We have summarized our experience of using MRI with retrograde contrast enhancement for diagnostics of rectovaginal fistulas.

Journal of Physics: Conference Series **677** (2016) 012011 doi:10.1088/1742-6596/677/1/012011

2. Methods

a. MRI acquisition
MRI was performed with MAGNETOM ESSENZA scanner (SIEMENS, Germany), with field strength of 1.5 T, using a body matrix coil in T2-weighted sequence. Both axial and sagittal projections were used, with a section thickness of 3 mm (TR-4900 ms, TE-87 ms). The examination was performed after a standard bowel preparation. MRI was conducted prior to and after retrograde filling of the rectum. In all cases, the doze administered through the rectum was well tolerated by patients.

2.2 Modification of MRI techniques
We filled the rectum with 100-120 ml of a radiopaque to make MRI more informative. We used a special ultrasound gel as a radiopaque to be filled in the rectum.

3. Clinical example

b. Clinical example 1
Patient M., female, 50 years of age, received the combination therapy of rectal cancer (stage IIA, with pT3N0M0) in our clinic. The neoadjuvant radiation therapy (total dose 40 Gy) and the anterior resection of rectum were performed. After the surgery, a rectovaginal fistula developed. Gynecological and ultrasound examinations failed to detect the fistula. MRI failed to detect the fistula until contrast enhancement of the rectum with the ultrasound gel was used (Figure 1).

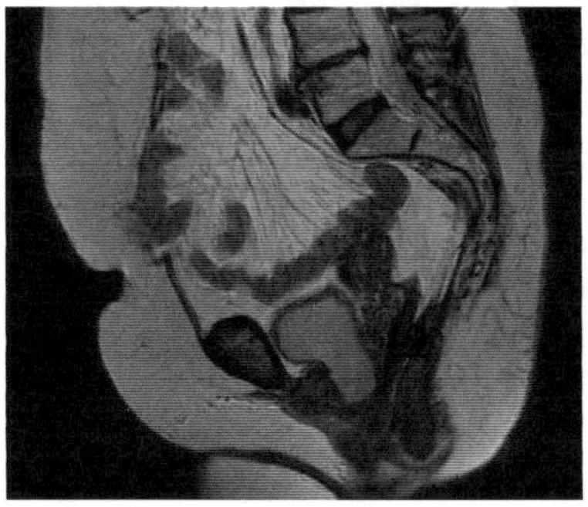

Figure 1. Sagittal T2-weighted MRI images without contrast enhancement of the rectum with the ultrasound gel. No fistula is identified.

Contrast enhancement of the rectum with 120 ml of the ultrasound gel allowed us to visualize the rectum and the vagina filled with the radiopaque and to identify the fistula on the anterior wall of the rectum. (Figure 2).

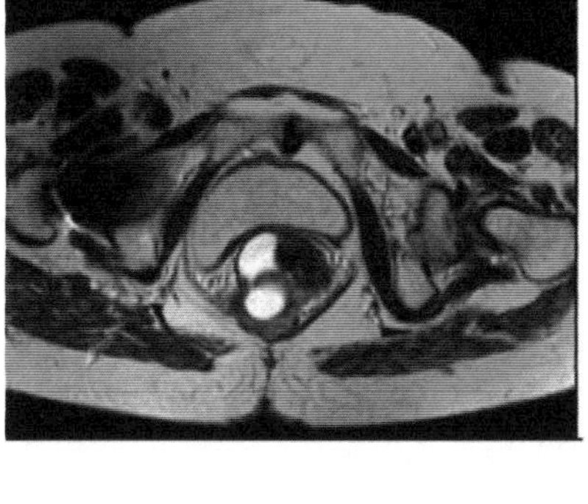

Figure 2. Axial T2-weighted MRI images with contrast enhancement of the rectum with the ultrasound gel. A fistula, as well as the rectum and the vagina filled with the gel, are identified.

The plastic reconstruction of the rectovaginal fistula was performed. At the expiration of 2 months after the surgery, MRI without contrast enhancement of the rectum did not identify a fistula (Fig. 3a, 3b). Contrast enhancement of the rectum with the gel did not show a fistula either. The gel did not penetrate into the vagina. (Figure 3).

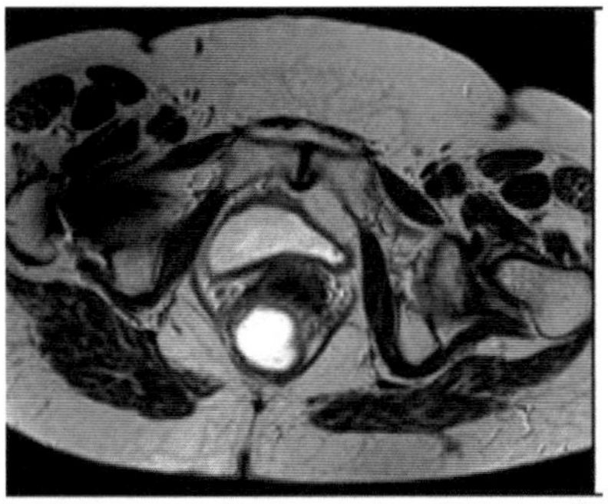

Figure 3. Axial T2-weighted MRI images with contrast enhancement of the rectum with the ultrasound gel. No fistula is identified.

c. Clinical example 2

Patient G., female, 63 years of age, received the therapy of rectal cancer (stage IIA, with pT3N0M0) in our clinic. The neoadjuvant radiation therapy (total dose 40 Gy) and the anterior resection of rectum were performed. After the surgery, a rectovaginal fistula was identified. The fistula was not detected by gynecological examination. Video colonoscopy identified a 2-3 mm opening on the anterior wall of the rectum. When air was supplied, the patient noted some air coming out from her vagina. The plastic reconstruction of the rectovaginal fistula was performed. At the expiration of 3 months after the surgery, gynecological examination did not detect a fistula. Video colonoscopy identified a 1-2 mm opening on the anterior wall of the rectum.

MRI without contrast enhancement of the rectum did not visualize a fistula, but detected some signs of inflammation in the region of anastomosis (Figure 4). MRI with contrast enhancement of the rectum (120 ml of the gel) identified a rectovaginal fistula and showed a penetration of the gel into the vaginal lumen. (Figure 5).

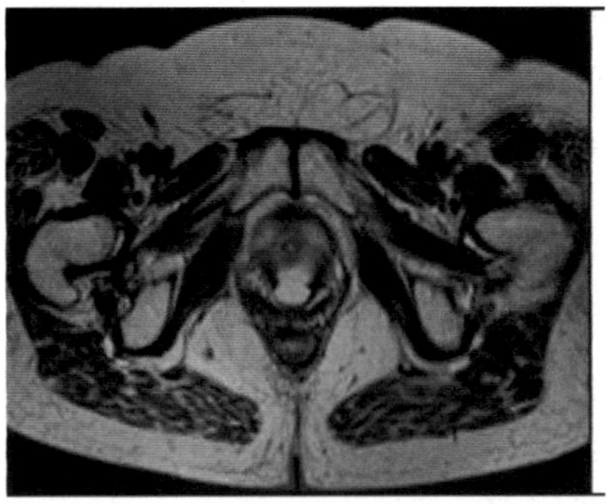

Figure 4. Axial T2-weighted MRI images without and with contrast enhancement of the rectum with the ultrasound gel. No fistula is identified

Taking into account the changes shown by MRI, the anti-inflammatory therapy was prescribed to the patient. Plastic reconstruction of the fistula is planned.

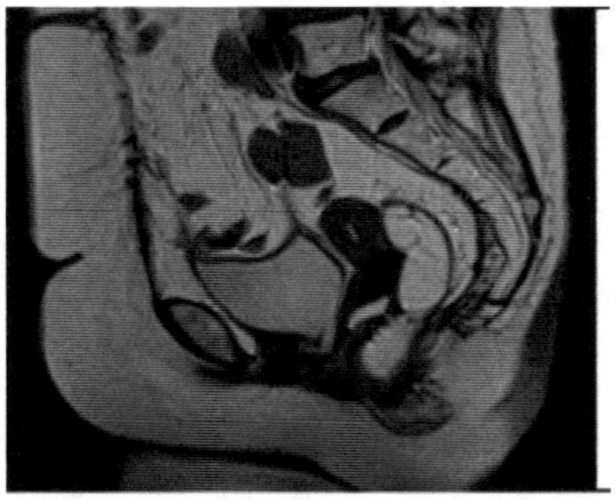

Figure 5. Sagittal T2-weighted MRI images with contrast enhancement of the rectum with the ultrasound gel. A fistula, as well as the rectum and the vagina filled with the gel, are identified.

4. Discussion

The presented case studies demonstrate that MRI with contrast enhancement of the rectum is informative for diagnostics of rectovaginal fistulas after combination therapy of rectal cancer. This method is informative when gynecological and ultrasound examinations fail. Fiber-optic colonoscopy of rectovaginal fistulas is effective, but it fails to evaluate the surrounding tissues and the rectal wall. The advantages of MRI with contrast enhancement of the rectal lumen include simplicity and a good tolerability. This method makes it possible to identify rectovaginal fistulas and to evaluate the rectal wall and the vaginal wall, which is useful for planning a surgical treatment of fistulas.

Acknowledgments

Authors acknowledge Russian Science Foundation (project № 14-45-00040) for financial support of the preparation of the manuscript.

References

[1] Watanabe J, Ota M, Kawaguchi D, et al 2015 *Int J Colorectal Dis* **8** 7
[2] Aliev SA 1999 *Vestnik surgery* **3** 66-70
[3] Brusov PG Inoyatov IM, 1996 *Surgery* **2** 45-48
[4] Petrov VP 2001 *Vestnik surgery* **6** 59-64
[5] Watanabe J et al 2015 *Int J Colorectal Dis* **8** 7
[6] Pan H, Wang L, Peng Y, Gu J 2015 *Zhonghua Wei Chang Wai Ke Za Zhil* **8(7)**656-60
[7] Brown G, Daniels IR, Richardson C et al 2005 *Br. J. Radiol* **78 (927)** 245–251
[8] Chun HK, Choi D, Kim MJ et al 2006 *AJR* **187 (6)**1557–1562
[9] Iafrate F, Laghi A, Paolantonio P et al 2010 *RadioGraphics* **26(3)**701–714
[10] Akasu T, Iinuma G, Takawa M, 2009 *Ann Surg Oncol* **16** 2787-2794
[11] Zhang XM, Zhang HL, Yu D, et al 2008 *AJR Am J Roentgenol* **190** 1271–1278
[12] Krestin GP, Steinbrich W, Friedmann G 1988 *Radiology.***168(2)**307-11
[13] Miucin-Vukadinović I, Kozić D, Adić O 2008 *Med Pregl* **3-4** 157-63
[14] Danko NA Vazhenin AV Vaganov NV 2011 *Siberian Journal of Oncology* **6 (48)**57-61
[15] Hoeffel C, Arrivé L, Mourra N, Azizi L 2011 *Radiographics.* **26(5)**1391-407
[16] Berezovskaya TP, Rukhadze GO, Nevolskih AA Silanteva NK 2014 *Coloproctology* **53** 52-53

Journal of Physics: Conference Series **677** (2016) 012012 doi:10.1088/1742-6596/677/1/012012

The role of MRI in the diagnosis of acute radiation reaction in breast cancer patient

Zh A Startseva[1,2], L I Musabaeva[1], AV Usova[1,3], I G Frolova[1], K A Simonov[1,2], V V Velikaya[1,2]

[1] Cancer Research Institute, Siberian Branch of the Russian Academy of Medical Sciences, Tomsk, Russian Federation
[2]National Research Tomsk Polytechnic University, Tomsk, Russian Federation
[3]Neurobiology Lab, Research Institute of Biology and Biophysics, Tomsk State University, Tomsk, Russian Federation

E-mail: simonov_ka@bk.ru

Abstract. A clinical case with acute radiation reaction of the left breast after organ-preserving surgery with 10 Gy IORT (24.8 Gy) conventional radiation therapy has been presented. Comprehensive MRI examination showed signs of radiation- induced damage to skin, soft tissues and vessels of the residual breast.

1. Introduction

Radiotherapy is important part of combined treatment of breast cancer. The outcomes of many randomized trials suggest that the use of postoperative radiotherapy significantly reduces the number of breast cancer recurrence, the overall 5 - year survival rate reaches high levels (87-95%) [3, 5, 6, 7].

Currently, a new method - intraoperative radiation therapy (IORT), which is conducted by an electron beam of different energy is an important direction in the radiation oncology [1, 2, 4]. The main advantage of IORT is that during the surgery the radiation source can be more accurately sum directly to the tumor bed.

In many centers after IORT performed distant gamma - therapy on the entire volume of the rest of the breast. One of the main problems in evaluation of the remaining breast tissue after intensive radiotherapy. Several, cases of radiation reactions and injuries after the operations and IORT breast tissue had been described in the literature [1, 2].

Tomsk Cancer Research Institute conducted a study of acute radiation reactions during the postoperative period. Typical reactions include erythema, dry and wet epidermatit and late radiation damage to the I-II severity. Magnetic - resonance imaging (MRI) is important characterization of radiation reactions and damage of the remaining breast tissue in conditions of high-dose radiotherapy.

2. Methods

A clinical case with acute radiation reaction of the left breast after organ-preserving surgery with 10 Gy IORT (24.8 Gy conventional radiation therapy) has been presented.

Patient Ch., age - 38 years old, was treated in the Tomsk Cancer Research Institute with the diagnosis of cancer of the left breast B II (T2N1M0). In December 2010, the patient was diagnosed: in the bottom - inner quadrant of the left breast was detected neoplasm 28x23x30 mm with irregular, indistinct contours, the presence of single calcifications. The density of high education, the structure of his uniform. Ultrasonography revealed a suspicion of metastatic lesion of the left axillary lymph node, which was confirmed by cytological examination after biopsy.

In January 2011, the patient performed organ-preserving surgery: sectoral resection of the left breast. Macropreparations: in breast tissue defined tumor site in diameter - 3 cm. After removal of the tumor with the capture of healthy tissue around the edge of a 1.0 cm by resected tumor bed held IORT 6 MeV fast electron, a single dose of 10 Gy, which corresponds to 24.8 Gy the standard rate of photon therapy.

The appearance of the left breast after healing wounds and stitches are removed 20 days after surgery and IORT of: the remaining part of the breast is greatly increased in size, swelling in the lower parts of the breast - marked flushing of the skin, radiation tissue infiltration in the area, where they spent the IORT, severe pain (figure 1).

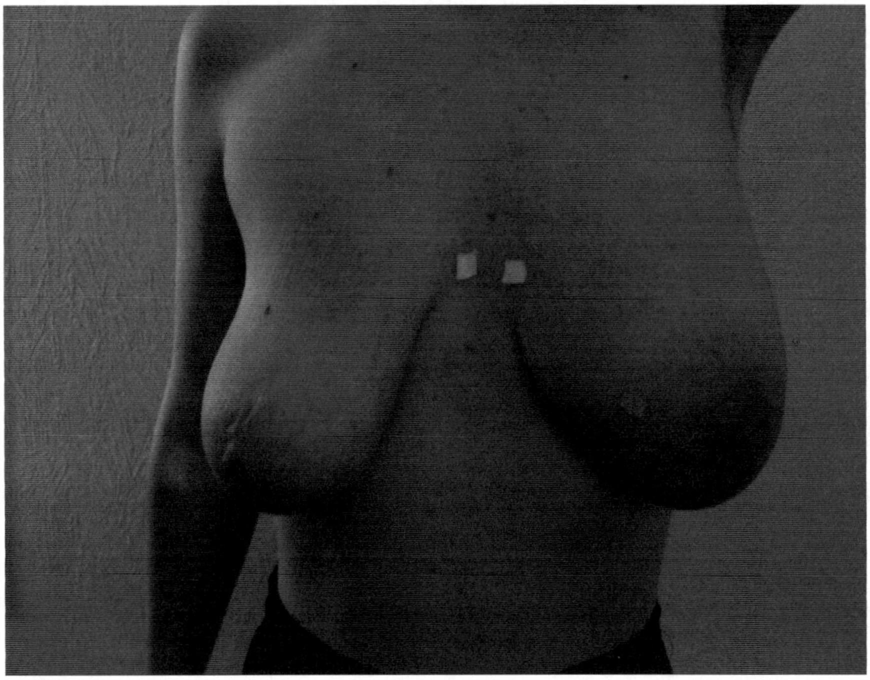

Figure 1. The appearance of the left breast by 20 days after surgery and IORT

To assess of the radiation reaction in the tissue of the left breast MRI study was conducted. According to the MR imaging of the mammary glands in T1-, T2-weighted mode and using dynamic contrast enhancement 10 ml Magnevist determined a significant asymmetry of the breasts by a massive edema of the left breast, and increase its size by more than half compared to the right. Are visualized postoperative changes of the left breast in the form of a unit cysts 2-3 mm fine hematoma 12 x 9 mm tissue swelling during the random access more pronounced (figure 2).

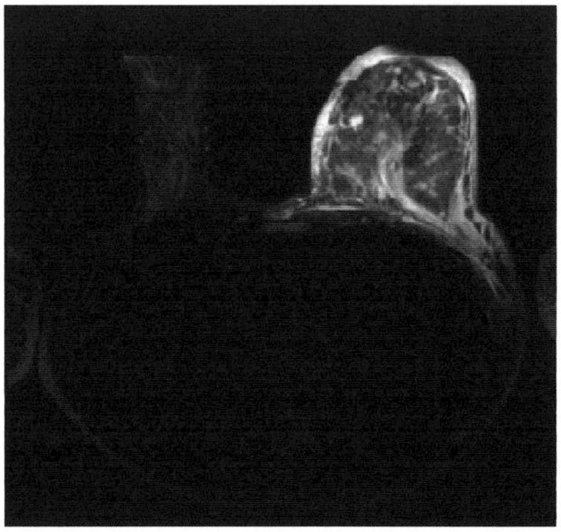

Figure 2. MRI: T2-TSE-Fat, the axial projection. Post-operative changes and intraoperative radiation therapy in a single dose – 10 Gy).

3D reconstructions subtraction scans arterial phase is determined by the asymmetry of blood flow to increase the diameter of blood vessels, and their number with the visualization of smaller branches, which is regarded as an expression of acute radiation reaction of the rest of the left breast after intraoperative radiation therapy (figure 3).

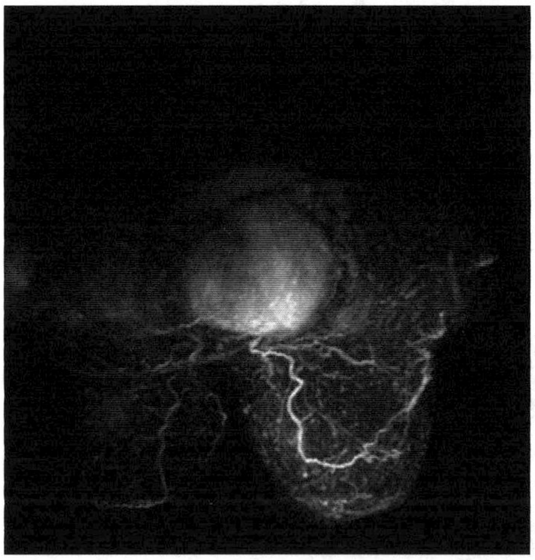

Figure 3. 3D reconstructions subtraction scans arterial phase. The course of laser therapy for treatment of acute radiation reaction: swelling and tenderness of the breast decreased significantly.

In the control MR imaging after laser therapy: positive dynamics - reducing the size of the left breast, swelling and severity of the reaction vessels (figure 4).

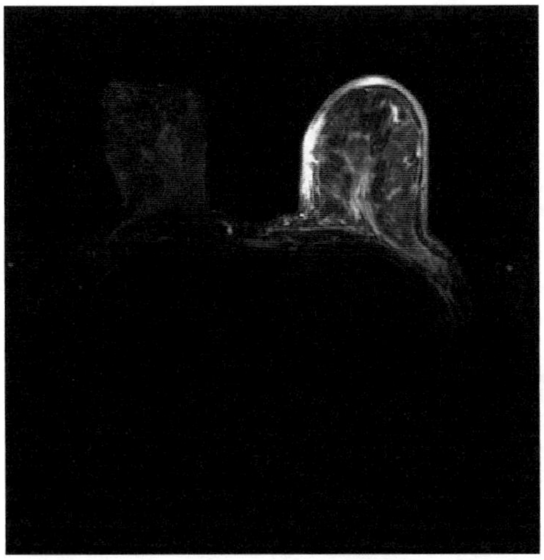

Figure 4. T2-TSE-Fat - sat, axial projection. Marked reduction in the size of the left breast, and the severity of tissue edema.

The patient was prescribed a course of external beam radiotherapy in standard dose fractionation for the rest of the breast. After reaching a total dose on the left breast - 20 Gy increased radiation reactionwas observed: bright redness of the skin of the left breast, hyperpigmentation, edema and the presence of a defect in the skin - more than 1 cm in the mammary fold as a manifestation of focal moist dermatitis. Remote gamma therapy has been interrupted for a week and continued magnetic laser therapy. A total of 15 treatments with a pronounced effect. The local dose of radiation therapy increased to 46 Gy. The total (IORT and RT) was 60 izoGy in the area of the bed of the removed tumor. In the area of regional lymph nodes - 44 Gy.

3. Conclusions

Thus, in the present case after breast-conserving surgery with 10 Gy (24.8 Gy) IORT diagnosed acute radiation reaction of the remaining breast tissue, the symptoms of which have been identified and described in detail MRI examination. The clinical picture of acute radiation reaction gradually changed after treatment: marked reduction in the size of the breast, the remaining breast tissue is softer, no pain, the patient's condition improved significantly.

References

[1] Medvedev F V, Mardinskyi Yu S, Gulidov I A, Smirnova I A 2010 *Oncosurgery* **2 (3)** 46
[2] Musabaeva L I, Zhogina Zh A, Lisin V A, Slonimskaya E M 2010 *Medical Radiology and Radiation Safety* **55 (5)** 42
[3] Musabaeva L I, Slonimskaya E M, Zhogina Zh A 2002 Manual for physicians Tomsk
[4] Musabaeva L I, Startseva Zh A, 2011 *Siberian journal of Oncology* **(2)** 79
[5] Simonov K A, Startseva Zh A, Slonimskaya E M 2013 *Siberian journal of Oncology* **2 (56)**
[6] Simonov K A, Startseva Zh A, Slonimskaya E M 2012 *The Siberian Scientific Medical Journal* **32(6)** 34
[7] Startseva Zh A, Simonov K A, Musabaeva L I, Velikaya V V 2013 *Radiation and Risk* **22 (3)** 72

Magnetic Resonance (MR) rectography in diagnostics of small-size rectal neoplasms

AV Usova[1,2], IG Frolova [1], NG Trukhacheva[1], OV Cheremisina[1], SG Afanas'ev[1]

[1]Cancer Research Institute, Siberian Branch of the Russian Academy of Medical Sciences, Tomsk 5, Kooperativny Street, Tomsk, Russia.

[2] Neurobiology Lab, Research Institute of Biology and Biophysics, Tomsk State University, Tomsk, Russian Federation

E-mail: afina.tsk@gmail.com

Abstract. Purpose was the assessment of diagnostic efficiency of MR-rectography in diagnostics of small-size rectal neoplasms. 12 patients with polyps and small tumors of a rectum are examined, the size of detected neoplasms varied in the range 3–18 mm. Native MRI and MRI with retrograde contrasting by ultrasonic gel was carried out. Results of MRI are compared with results of videocolonoscopy. Sensitivity of native MRT was 24%, MR-rectography was 88%. MR-rectography can be used in diagnostics of small-size rectal neoplasms.

1. Introduction

Colorectal cancer (CRC) belongs to the category of socially significant neoplasms. Nowadays CRC in Russia occupies the second place among malignant tumors in mortality. The prevalence of colorectal cancer in Russia in 2013 was 93.6 per 100 000 people [1]. According to the published data, the main part of patients now admit to specialized hospitals with the locally and regionally advanced process (T3-T4, N2-N3), and patients with III–IV stages are operated most often [2, 3]. The five-year survival rate of CRC patients with I–II stages is up to 90%, whereas the relapse rate at the locally advanced cancer achieves 46%, while the five-year survival rate is no higher than 38% [4]. In 60–90% cases, CRC arises in "adenomatous polyp–cancer" series [5]. The rate of cases of CRC development from villous adenomas (polyps) greater than 2cm in diameter is 35-53%, while for polyps greater than 3cm in diameter the probability of their malignization is 100%. Villous tumors are observed most often in rectum (80%) and, to a lesser degree, in the rectosigmoid section and sigmoid colon [6, 7, 8]. Thus, the problem of early diagnostics of small-size volume neoplasms in rectum is an important task in prevention of cancer, and the early detection of tumors will increase the survival rate and improve the late fate.

The leading methods in diagnostics of colorectal polyps and tumors are endoscopic methods, such as videocolonoscopy (VCS). The use of modern endoscopic instrumentation and novel methods, such as chromocolonoscopy, narrow-band, magnifying, and autofluorescence endoscopy, allows the high-accuracy visualization of minimal pathological changes in colon [9, 10]. An important advantage of the endoscopic method is the possibility of sampling for morphological examination. The VCS sensitivity ranges from 75 to 95%, and specificity achieves 98% [11, 12]. One of VCS disadvantages is relative intolerance due to pronounced pain syndrome, which not always can be controlled by local anesthesia. Examination under general anesthesia is not always possible because of technical impossibility and for patients with contraindications to general anesthesia, which leads to withholding of examination.

The relatively new method of examination is virtual colonoscopy (VCS) – method of noninvasive colon examination based on indirect visualization from data of X-ray computer tomography. The VCS sensitivity in detection of colorectal polyps is 93%, and specificity is 78% [13, 14, 15]. The diagnostic efficiency of this method is inversely proportional to the size of neoplasm. In particular, the sensitivity of this method in diagnostics is 90% for polyps greater than 1cm, 80% for polyps of 0.5–0.9 cm, and 67% for polyps no greater than 5mm [16, 17]. However, this method did not become a frequent routine practice due to the complex reconstruction algorithm and the need in colon distention, which is often poorly tolerable by patients. Also, virtual colonoscopy evaluates the internal relief of colon, while colon walls and surrounding structures can be evaluated only from data of native sections, whose tissue contrast is rather low.

MRI is a method of choice in diagnostics of pathologies in the small pelvis [18, 19, 20]. However, the efficiency of native study in diagnostics of rectal polyps and small-size tumors is limited due to the complex relief of the mucous tunic, close intensities of MR signals from polyps and from unchanged colon wall, as well as difficult visualization of small-size tumors against the background of unchanged colorectal mucous tunic.

A possible solution for this problem is MR colonography, which allows colon walls to be distended due to retrograde filling of rectum with contrasting agent and thus the method efficiency to be improved [21]. However, the data on the efficiency of this method at rectal polyposis and small-size neoplasms are few. In the study published by Purkayastha [22], the method sensitivity and specificity were 91% and 98%, respectively. As in the case of virtual colonoscopy, the diagnostic efficiency of the method is inversely proportional to the polyp size [23]. Disadvantages of MR colonography include the relatively long duration of the procedure, impossibility of high-accuracy examination all over the colon length, as well as presence of peristaltic artifacts caused by colon wall distension. Solutions used as intraluminal contrast agent also cause artifacts from movement in colon lumen. Artifacts from movement caused by patient discomfort at colon distension introduce additional difficulties.

Nowadays there is no technique combining high efficiency and good subjective tolerance for colorectal examination for polyposis and small-size neoplasms. All the above-said cause the need in MRI optimization in order to increase the MRI efficiency and to provide the subjective tolerance of examination in diagnosis of colorectal pathological changes.

2. Methods

2.1. Characteristics of patients

Twelve (seven male and five female) patients with colorectal polyps and small-size tumors (malignized polyp) were examined by the standard MRI method and by the modified technique (MR rectography).

The average age of the patients was 56.8 years. The patients with suspected volume colorectal neoplasms undergoing examination in our institute were included in the study.

2.2. MRI aqusition

MRI was carried out at the MAGNETOM ESSENZA (SIEMENS) 1.5T MR tomograph. The MRI protocol included T2-VI in three planes with section thickness of 3 mm, (TR-4900ms, TE-87ms).

2.3. Modification MRI techniques

In our study, we propose modification of standard MRI examination procedure with isolated retrograde filling of rectum with 120-150 ml of viscous contrast substance. We have chosen a gel for ultrasonic examination as a contrast agent owing to its availability and biological passivity. The volume injected per rectum in all cases was well tolerated by patients.

2.4. Evaluation of images

The size of detected neoplasms varied in the range 3–18 mm. Examination was conducted before and after the retrograde filling of rectum with determination of the number of neoplasms. The number of neoplasms detected from data of native MRI and with the use of MR rectography was compared with VCS data.

Criteria of the presence of a volume colorectal neoplasm from the data of MR rectography were local thickening or exophytic evagination of smoothened colon wall against the background of intraluminal contrast agent. Neoplasms had sharp, even or torous contours, the intensity of MR signal was close or equal to that from adjacent parts of unchanged colorectal mucous tunic, and in all the cases the external layer of colon remained sharp (Figure. 1).

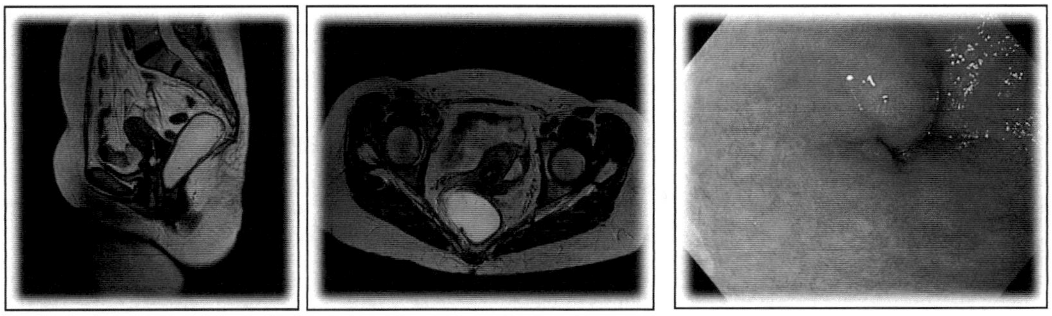

T2-weighted images T2-weighted images Video colonoscopy

Figure. 1. T2-weighted images of a rectal polyp in case of retrograde filling with a gel . Video colonoscopy of the same patient.

Differential diagnostics of malignant and nonmalignant neoplasms was not carried out.

3. Results

Results are summarized in Table 1.

Table 1. Results of native MRI and MR rectography in comparison with virtual colonoscopy (VCS).

patient	Number of neoplasms Standard MRI	truth positive	false negative	Number of neoplasms MRI rectography	truth positive	false negative	Number of neoplasms VCS
1	1	1	1	2	2	0	2
2	0	0	1	1	1	0	1
3	1	1	4	3	3	2	5
4	0	0	2	2	2	0	2
5	1	1	3	5	4	1	4
6	0	0	3	3	3	0	3
7	0	0	3	2	2	1	3
8	2	1	0	1	1	0	1
9	1	1	1	3	2	0	2
10	0	0	1	1	1	0	1
11	1	1	1	2	2	0	2
12	1	1	2	2	2	0	3
total	8	7	22	27	25	4	29

The sensitivity of native MRI in detection of small-size volume colorectal neoplasms was 24% (7/29) and 86% (25/29) with the use of MR rectography. At the native MRT, the number of false negative cases was 22 because of the difficult separation of small-size neoplasms against the background of colorectal folds. With MR rectography, four false-negative cases were detected, which was connected with the small size of neoplasms in three cases, and in one case a polyp was assigned as a near-wall gas bubble.

4 Discussion

MR colonography is highly sensitive for detection of rectal polyps (86%) and does not cause discomfort to patients. MR rectography is not positioned as an alternative to video colonoscopy. This method can be used for identifying polyps and small malignant tumors of the rectum, as well as for dynamic monitoring of patients who do not tolerate video colonoscopy or have contraindications to it.

References

[1] A Davydov MI, Aksel' EM 2010 *Vestnik RONTs im. N.N. Blokhina RAMN* **21 2 (80)** (Suppl. 1)
[2] Butenko AV, Razbirin VN. 2011 *Sib. Onkolog. Zhurnal.* **6 (48)** 83–89
[3] State of cancer care in Russia in 2011. Ed. by VI Chissov, VV Starinskii, GV Petrova. Moscow, 2012. 240
[4] Berlin JW, Gore RM, Yaghmai V, Newmark GM 2009 *Semin Roentgenol.* **35** 370–384
[5] Laoteva EA, Kozlova IV, Myalina YuN. 2013 *Saratov Nauchno-Med. Zhurnal.* **9** 2
[6] Egorenkov VV, Moiseenko FV 2010 *Prakt. Onkologiya* **11** 81–87
[7] Shul'pekova you 2002 *Ros. Zhurn. Gastroenterol. Gepatol. Koloproktol.* **12(4)**64–68
[8] Zinkiewicz K 2004 *Dis. Colon. Rectum.* **47**1115
[9] Kashin SV, Zav'yalov DV, Kamkina GV, Akhapkin NV 2012 *Klinik. Edoskopiya* **34** 16-25
[10] The Paris endoscopic classification of superficial neoplastic lesions: esophagus, stomach and colon: November 30 to December 1, 2002. *Gastrointest Endos* **58** 3-43
[11] Jover R.et al. 2012 WEO *Colorectal Cancer Screening Committee Meeting* **5(18)** 23
[12] Matsuda T 2012 *WEO Colorectal Cancer Screening Committee Meeting* **5(18)**54
[13] Pickhardt PJ, Hassan C, Halligan S, Marmo R 2011 *Radiology* **259** 393–405
[14] Atkin W, Dadswell E, Wooldrage K 2013 *Lancet* **381** 1194–202
[15] Halligan S, Wooldrage K, Dadswell E 2013 *Lancet* **381** 1185–93
[16] de Haan MC, van Gelder RE, Graser A, Bipat S, Stoker J 2011 *Eur Radiol* **21** 1747–63
[17] White TJ, Avery GR, Kennan N et al. 2009 *Corectal. Disease* **11(2)**138–145
[18] Prorokov VV, Malikhov AG, Knysh VI. 2002 *Prakt. Onkologiya.* **3(2)**1–5
[19] Brown G, Daniels IR, Richardson C et al. 2005 *Br. J. Radiol.* **78 (927)** 245–251
[20] Chun HK, Choi D, Kim MJ et al. 2006 *AJR.* **187 (6)** 1557–1562
[21] Laghi A, Bellini D, Petrozza V et al 2015. *Colorectal Dis.* **1(17)**36-43
[22] Purkayastha S, Athanasiou T, Tekkis PP, Constantinides V 2007 *Colorectal Dis.* **9** 100–110
[23] Zijta FM, Bipat S, Stoker J 2010 *Eur Radiol* **20** 1031–46